P9-BIY-259

Student Resource Manual
to Accompany

Differential Equations

A Modeling Perspective

Robert L. Borrelli
Courtney S. Coleman

•

Harvey Mudd College

John Wiley & Sons, Inc.

New York • Chichester • Weinheim • Brisbane • Singapore • Toronto

COVER PHOTO   © Richard Megna/Fundamental Photographs

Copyright © 1998 by John Wiley & Sons, Inc.

Excerpts from this work may be reproduced by instructors for distribution on a not-for-profit basis for testing or instructional purposes only to students enrolled in courses for which the textbook has been adopted. *Any other reproduction or translation of this work beyond that permitted by Sections 107 or 108 of the 1976 United States Copyright Act without the permission of the copyright owner is unlawful. Requests for permission or further information should be addressed to the Permissions Department, John Wiley & Sons, Inc., 605 Third Avenue, New York, NY 10158-0012.*

ISBN 0-471-24589-5

Printed in the United States of America

10 9 8 7 6 5 4 3 2 1

Printed and bound by Hamilton Printing Company

Student Resource Manual
to Accompany

Differential Equations

A Modeling Perspective

To use or not to use: that is the question:
Whether 'tis nobler in the mind to suffer
The stings and errors of outrageous computation,
Or to apply technology against a sea of questions,
And by its power resolve them?
To model: perchance to visualize: ay, there's the key:
For in that model what insights may come...

Preface

The problems in each section of the text are arranged roughly in order of increasing challenge. The beginning problems in each set are usually tied to specific examples worked out in the text, and they can be solved by following the outline of one of the examples. This manual has solutions (including graphs) for every other part of every odd-numbered problem. Your instructor will expect you to write your own solutions, perhaps based on what you see here.

Getting the Most out of the Problems

- Look over the assigned problems and identify the related examples in the text.

- Try to do the problems on your own, but if you aren't successful then you may want to look at solutions of similar problems.

- Explain your answers, how you got them, and why they provide solutions of the problems.

- The figures in the manual give you some idea of what your own figures should look like.

- Put enough detail in your figures to be informative, but not so much that it is confusing. Use the graphs in the text and this manual as your guides. To improve the appearance of your figures you may want to change the scales on the axes, the solve time, or the initial data.

- Explain how your figures answer the questions posed in the problem.

Getting the Most Out Of Your ODE Solver

- Check out the numerical solvers available to you, and practice using them on simple ODEs. Explore the capabilities and limitations of your solver by applying it to some of the figures in Chapter 1 of the text.

- Some features in the text and manual figures may not be available on your numerical solver. If that happens, check with your instructor about ways to work around the difficulties.

- In some cases a CAS (Computer Algebra System) can be used to get a solution formula, which can then be graphed.

Getting the Most Out of a Group Project
•

- A group project requires the efforts of a team so that it can be completed in a reasonable length of time.

- The way you will work on a group project is different from the way you work on an individual problem. Communication within the group is critical in order to stay focused on the goal. Be open to new ideas and approaches, but at the same time be honest in expressing your opinions and evaluations of your colleagues' suggestions.

- The write-up of your group project is expected to be longer and more thoroughly documented than the usual write-up of an individual problem, so be prepared to spend more time.

Acknowledgments
•

We particularly want to thank the students at Harvey Mudd and Pomona Colleges, Arizona State University, and several other colleges and universities who have helped us get the problems and the solutions right. Any mistakes in the manual, though, are ours, and we would appreciate it if you send in your corrections and suggestions.

Special thanks go to the following Harvey Mudd College and Claremont Graduate University students (current and former) for all they have done to help with the text, problems, solutions, graphics, and supporting software: Aaron Archer, Tiffany Arnal, Kevin Carosso, Patri Forwalter-Friedman, Ned Freed, Motoya Kohtani, Aaron Lamb, Claire Launay, Christie Lee, Tony Leneis, Dan Lopez, Susan McMains, Joel Miller, Dan Newman, Robert Prestegard, Justin Radick, Dave Richards, Itai Seggev, Marie Snipes, Treasa Sweak, Jenny Switkes, Kal Wong, Xuemei Wu, Kaiqi Xiong, and Bob Zirpoli.

A Final Word
•

Keep an open mind while doing your assignment, and have fun along the way using your computer. If you have comments or suggestions about anything in this manual or in the text, send e-mail to John Wiley & Sons at math@jwiley.com.

R. L. Borrelli
C. S. Coleman
August 1997

Contents

4 • *Applications of Second-Order Differential Equations*   **77**

5 • *Systems of Differential Equations*   **85**

6 • *The Laplace Transform*   **97**

7 • *Linear Systems of Differential Equations*   **110**

8 • *Stability*   **137**

First-Order Differential Equations and Models

The main thrust of this chapter is to find solution formulas for first-order ODEs, discuss some ODE models, introduce the geometry of solution curves, and show the central role of numerical solvers in understanding the behavior of solutions. Solution methods are tabulated at the end of the chapter for reference purposes. Questions of the existence and uniqueness of solutions of initial value problems (IVPs) and the sensitivity of IVPs to changes in the data can be answered from these formulas. The relationships between ODEs and dynamic processes are so basic that we explore the connections right from the start. Throughout the chapter, simple processes are introduced and modeled with ODEs. We hope users of this text will use numerical solvers to approximate and graph solutions of ODEs; the computer icon designates problems for which a numerical solver is appropriate. Most CAS packages have numerical solvers, but can also produce solution formulas for many first-order ODEs (even ODEs with parameters), not to mention graphs and direction fields.

Systems of ODEs are introduced in this chapter for several reasons. Systems arise naturally in modeling. Some systems can be solved using the first-order techniques of the chapter. All numerical ODE solvers available on the market today are designed to solve systems of first-order ODEs.

Note that discontinuous driving terms (e.g., the Engineering Functions of Appendix B.1) are introduced early in a mild way and in a more substantial way later in a modeling environment (Section 1.8). If your solver does not have these functions as predefined functions, you can often find a work-around using if-then-else logical commands over a finite time interval.

1.1 A Modeling Adventure

Comments

In this introductory section we hit upon several of the main themes of our approach to modeling and differential equations. We construct an initial value problem that models a changing fish population. The reasoning process leading to that IVP is called *modeling* and the resulting IVP is called a *model*. In the figures we visualize the long-term behavior of solution curves of some of the differential equations used in the models. We examine the "sensitivity" of the structure of these solution curves as the initial value y_0 or the harvesting rate H is changed, although we don't yet explicitly use the term "sensitivity." This scenario is one that will be repeated many times in this book. So if you didn't grasp everything this first time around, just wait for the next opportunity. As in any field, there are a lot of terms to be defined. We have opted to include some of the definitions in this opening section, but postpone others till later.

Tips on using numerical solvers

Like any other tool, you will need a "break-in" period to use your numerical solver effectively. The documentation that comes with the solver is a good place to start, but there is no substitute for hands-on experience. Some solvers are command-line driven with a precise (and often complex) syntax for entering commands, but many solvers have a graphical interface which allows the user to select items from a menu and makes it easy to enter data and display graphs. Our advice is to get started right away in using your solver, even before you completely understand the basic concepts behind the differential equations you are studying. The reason is that the graphical displays created by the solver will give you a better feeling for solutions of ODEs and how they behave when the underlying model changes.

From time to time we will give practical tips that will help you avoid some of the frustrations in using your solver. Here is some advice you can use right now:

- Numerical solvers are designed to solve initial value problems where the ODE is in normal form. Before using your solver, write your ODE in normal form and identify the rate function. Next, examine the initial condition and identify the initial time t_0 and the initial value y_0. Finally, choose a solve-time interval for either forward or backward solving. Your solver may ask you for these items. Some solvers allow users the option to enter initial points graphically by clicking directly on the screen. Most solvers allow users to insert settings for the solver itself, but these are often set by default and not adjusted until something goes wrong.

- If your solver does not give you the option of setting your own scale on the axes at the outset, then not to worry: go ahead and let your solver set the scales automatically and then look at the result. It may be just what you wanted. If not, then at this stage your solver will let you select your own scales on the axes.

- Your solver may continue to crank away and never seem to come to an end. In that case you may want to abort and take a shorter solve-time interval. This usually happens when rate functions are very large, and so examining the rate function at the outset will give you a clue that you may run into a problem with the choice of a solve-time interval.

- Many solvers allow the user to choose the number of computed points (equally spaced in time) to display on the screen; these points are then connected with straight line segments. If the graph produced by your solver looks partly like a broken-line graph, then you should go back and increase the number of displayed points and solve again.

- Some solvers have trouble dealing with IVPs like

$$y' = ay - cy^2 - H(t), \qquad y(t_0) = y_0$$

if the harvesting function H is not continuous (for example, a step function). One way to cope with the problem is to choose a very large number of displayed points, but this doesn't work with all solvers. So you may have trouble reproducing the figures in Problems 6 and 7. In any case, your solver will have to work hard (or be clever) in order to handle on-off harvesting functions (Appendix B.1 describes these functions). Here's a way that some solvers deal with this problem: The solver computes values at the equally spaced time points displayed on the computer screen by using internal time-steps which are selected automatically in order to stay within specified error bounds (selected by default), but even this process can be defeated by discontinuous harvesting functions. The problem appears to be that internal time-steps become so large that they miss the on-off points of the harvesting function. Some solvers have a setting that allows the user to select a maximum internal step size, and if it is set low enough then the solver will be better able to "see" the on-off points of the harvesting function. Some really sophisticated solvers will handle this fairly well when on-off functions are detected.

---

1. **(a).** You can solve the IVP, $y' = ay$, $y(0) = y_0$, the easy way by replacing H by 0 in formula (9), the formula that gives the solution of initial value problem (4). So $y = y_0 e^{at}$ for $t \geq 0$. We make the restriction $t \geq 0$, not because there is any difficulty in defining the exponential for negative values of t, but because the fishing model has been formulated for future projections only. If the initial value problem (4) is valid for the past as well, then we don't need the restriction $t \geq 0$.

   You could also repeat the steps given in this section for getting from IVP (4) to formula (9), replacing H by 0 every step of the way. This step-by-step derivation is an aid to understanding just why the formula $y = y_0 e^{at}$ gives the unique solution for each value of y_0. This solution formula makes sense if you remember from calculus that the exponential function e^{at} is the only function whose derivative is a multiple of itself.

3. **(a).** The terms in the model ODE, $y' = ay - cy^2 + R$, have the following interpretation: $y(t)$ is the fish tonnage at time t, y' is the rate of change $dy(t)/dt$ in the fish tonnage; a is the difference between the population's birth and death rates (ignoring both overcrowding and harvesting); c is the overcrowding coefficient and is measured in $(\text{ton·year})^{-1}$ so that cy^2 has units of tons/year; R is the restocking rate in tons/year and is assumed to be a positive function, but not necessarily constant.

5. **(a).** The model IVP, $y' = ay - cy^2 - H_0 y$, $y(0) = y_0$, where a, c, H_0, and y_0 are positive constants, models a harvesting situation where the harvesting rate at time t, $H_0 y(t)$, is proportional to the tonnage $y(t)$ of fish. The term y' represents the change in fish tonnage y; a is the difference between the population's birth and death rates (ignoring overcrowding and harvesting); c is the overcrowding coefficient. The harvesting rate $H_0 y(t)$ makes a lot of sense because it means that the tonnage caught rises or falls with the amount of fish actually in the lake. This is called a constant effort model because the ratio of the harvesting rate to the population is the constant H_0, so the same amount of effort is expended whatever the size of the population. One day your nets will be full because there are a lot of fish in the

lake, another day with the same effort your nets will be nearly empty because the number of fish is low.

7. Our solver uses the square wave function $A \, \text{sqw}(t, d, T)$ to represent a periodic on-off function with period T, amplitude A, and "duty cycle" $d\%$. Duty cycle is an engineering term for the percentage of the total period that the function is actually on; it is assumed that the on-time is always at the beginning of each period of the cycle. For example, $4 \, \text{sqw}(t, 50, 1)$ represents an on-off function of amplitude 4 that has value 4 for the first half (i.e., 50%) of the cycle of period 1, and value 0 for the last half. See Appendix B.1 for graphs of $\text{sqw}(t, d, T)$ and other on-off functions that are widely used in engineering, science, and mathematics.

Careful measurements of the two graphs in Figures 1.1.7 and 1.1.8 show that harvesting is "on" for the first two months of each year (since the solution curves fall during these periods) in the first graph (so $d = 100/6 = 16.66\%$), and "on" for the first eight months of the year in the second graph (since now the curves fall during the first 8/12 of each year) [so $d = (8/12)100 = 66.66\%$]. The graphs are similar, but the initial population size needed for survival starts a lot lower (somewhere slightly below $y_0 = 1.0$) in the two-month harvest season case. With an eight-month season there is a dangerously wide extinction range of values of y_0, extending all the way up to somewhere between $y_0 = 4.5$ and $y_0 = 5$.

Here's a **computer tip**: if possible, lower the maximum internal step size of your numerical solver so that your solver doesn't overlook the on-off times of the harvest season. We lowered the maximum internal step size of our solver from 1.0 to 0.05 to avoid this problem.

Problem 7, Graph 1.   Problem 7, Graph 2.

Background Material: Growth Processes

The U.S. Bureau of the Census frequently predicts population trends. To do this, demographers use known population data to formulate "laws" of population change. Each law is converted into a mathematical formula that can then be used to make the predictions. Since there is a good deal of uncertainty about which law best describes actual population shifts, several are derived, each proceeding from different assumptions. Similarly, ecologists make predictions about changes in the numbers of fish and animal populations, and biologists formulate laws for changing densities of bacteria growing in cultures. Let's introduce some general laws of population change, solve the corresponding mathematical models, and interpret the results in terms of the growth, stabilization, or decline of a population.

Background Material: Growth Models

Suppose that $y(t)$ denotes the population at time t of a species. The values of $y(t)$ are integers and change by integer amounts as time goes on. However, for a large population an increase by one or two over a short time span is infinitesimal relative to the total, and we may think of the population as changing continuously instead of by discrete jumps. Once we assume that $y(t)$ is continuous, we might as well smooth off any corners on the graph of $y(t)$ and assume that the function is differentiable. If we had let $y(t)$ denote the population *density* (i.e., the number per unit area or volume of habitat), the continuity and differentiability of $y(t)$ would have seemed more natural. However, we shall continue to interpret $y(t)$ as the size of the population, rather than density.

The underlying principle

$$\text{Net rate of change} = \text{Rate in} - \text{Rate out} \tag{i}$$

applies to the changing population $y(t)$. For a population, the "Rate in" term is the sum of the birth and the immigration rates, while the "Rate out" term is the sum of the death and the emigration rates. Let's regroup the rates into an internal rate (birth minus death) and external rate (immigration minus emigration). Averaged over all classes of age, sex, and fertility, a typical individual makes a net contribution R to the internal rate of change. The internal rate of change at time t is, then, $Ry(t)$, where

$$Ry = (\text{Individual's contribution}) \times (\text{Number of individuals})$$

The *intrinsic rate coefficient* R will differ from species to species, but it always denotes the average individual's contribution to the rate. Given the rate coefficient R and the migration rate M, we see that

$$y'(t) = Ry(t) + M \tag{ii}$$

and the study of population changes becomes the problem of solving ODE (ii) for $y(t)$.

In many cases R and M can be approximated by constants or else by simple functions of the population levels and time. If simple models suffice for observed growth processes, there is little need for more complex assumptions. In modeling, it's considered better to have fewer assumptions.

Background Material: Exponential Growth

It may safely be pronounced, therefore, that population, when unchecked, goes on doubling itself every twenty-five years.

Malthus

Thomas Robert Malthus (1766–1834) was a professor of history and political economy in England. The quotation is from "An Essay on the Principle of Population As It Affects the

Future Improvement of Society." Malthus's views have had a profound effect on Western thought. Both Darwin and Wallace have said that it was reading Malthus that led them to the theory of evolution.

The Malthusian principle of explosive growth of human populations has become one of the classic laws of population change. The principle follows directly from (ii) if we set $M = 0$ and let $R = r$, a positive constant whose units are $(\text{time})^{-1}$. ODE (ii) becomes the linear ODE $y' = ry$ which has the exponentially growing solution

$$y(t) = y_0 e^{rt}, \quad \text{for all } t \geq 0 \tag{i}$$

where y_0 is the population at the time $t = 0$. We see from (i) that the *doubling time* of a species is given by $T = (\ln 2)/r$ since if $y(t + T) = 2y(t)$, then

$$y_0 e^{r(t+T)} = 2y_0 e^{rt}$$

$$e^{rT} = 2$$

$$T = \frac{1}{r} \ln 2$$

Note the close connection with the half-life of a radioactive element (see Section 1.5).

Malthus claimed a doubling time of 25 years for the human population, which implies that the corresponding rate coefficient $r = (1/T) \ln 2 = (1/25) \ln 2 \cong 0.02777$. Let's use this information to calculate the annual percentage increase in a population with a 25 year doubling time. The solution formula (i) implies that

$$y(t + 1)/y(t) = \frac{y_0 e^{r(t+1)}}{y_0 e^{rt}} = e^r$$

and so Malthus's value of $r = 0.02777$ gives us that $y(t + 1)/y(t) \approx 1.0282$. This corresponds to a 2.8% annual increase in population. Malthus's figure for r is too high for our late-twentieth-century world. However, individual countries, for example, Mexico and Sri Lanka, have intrinsic rate coefficients which exceed 0.02777 and may be as high as 0.033 with a corresponding doubling time of 21 years. You may want to use an Almanac to calculate the intrinsic rate coefficient for the U.S.A.

Background Material: Logistic Growth

> *The positive checks to population are extremely various and include ... all unwholesome occupations, severe labor and exposure to the seasons, extreme poverty, bad nursing of children, great towns, excesses of all kinds, the whole train of common diseases and epidemics, wars, plague, and famine.*
>
> *Malthus*

The unbridled growth of a population (as predicted by the simple Malthusian law of exponential increase) cannot continue forever. Malthus claimed that resources grow at most arithmetically, i.e., the net increase in resources each year does not exceed a fixed constant. An exponential increase in the size of a population must soon outstrip available resources. The resulting hardships would surely put a damper on growth.

The simplest way to model restricted growth with no net migration is to account for overcrowding by requiring the rate coefficient R to have the form $r_0 - r_1 y$, where r_0 and r_1 are positive constants. Since $r_0 - r_1 y$ is negative when y is large, $y' = (r_0 - r_1 y)y$ is negative and the population declines. It is customary to write the rate coefficient as $r(1 - y/K)$, where r and K are positive constants, rather than as $r_0 - r_1 y$. We then have the *logistic equation* with initial condition,

$$y' = r\left(1 - \frac{y}{K}\right)y, \qquad y(0) = y_0 \tag{i}$$

Observe that $y(t) = 0$ and $y(t) = K$ are constant solutions of the logistic equation, the so-called *equilibrium* solutions. The coefficient r is the *logistic rate coefficient* and K is the *carrying capacity* of the species.

The logistic equation and its solutions were introduced in the 1840s by the Belgian statistician Pierre-François Verhulst (1804–1849). He predicted that the population of Belgium would eventually level off at 9,500,000. The estimated 1994 population of a little more than 10,000,000 is remarkably close to the predicted value.

Logistic laws are well suited to laboratory or other isolated populations for which there is reason to believe that K and r are constants. Beginning with the experiments of the Soviet biologist G.F. Gauze in the early 1930s, there have been numerous experiments with colonies of protozoa growing under controlled laboratory conditions. The results of these experiments generally confirm the logistic model. See J.H. Vandermeer, "The Competitive Structure of Communities: An Experimental Approach with Protozoa," *Ecology* **50** (1969), pp. 362–371.

1.2 Visualizing Solution Curves

Comments

In this section we connect the form of the rate function $f(t, y)$ in the ODE $y' = f(t, y)$ to the behavior of a solution curve. For example, if $f(t, y)$ is positive at a point (t, y), then the solution curve through that point has a positive slope and is rising. The nullclines in the ty-plane divide the regions where solution curves rise from regions where they fall, so we discuss how to find nullclines by solving $f(t, y) = 0$. We discuss several other ways to visualize the behavior of solution curves.

---

1. The IVP is $y' = y - y^2/12 - H(t)$, $y(0) = y_0$, where

$$H(t) = \begin{cases} 4, & 0 \le t \le 5 \\ 0, & 5 \le t \le 10 \end{cases}$$

and y_0 lies between 0 and 20. The simplest way to find the smallest value of y_0 such that the population recovers after the harvesting stops is to set $y(5) = 0$, solve *backward* in t until $t = 0$, and then estimate the value of $y(0)$ from the graph of the approximate solution. See Fig. 1 where the points $(5, 0)$ and $(0, y(0))$ are specially marked. From the graph, $y(0) \approx 7.45$.

Problem 1.

3. The regions in R where a solution curve rises are precisely the regions where $y' > 0$. To find these regions, first rewrite each ODE in the form $y' = f(t, y)$. Then set $f(t, y) = 0$ so that we can find the equations of some parts of the boundary curve of the region where f is positive (i.e., where y' is positive). These boundary curves are the nullclines. In each graph we use a dashed line for the nullcline that defines the boundary.

(a). See Fig. 3(a). The region of rising solution curves for the ODE, $y' = 1 - y$, is the region below the horizontal nullcline line $y = 1$, (i.e., the region described by $y < 1$). Here the nullcline coincides with the equilibrium solution curve $y = 1$, so you can't see the dashed line.

(c). See Fig. 3(c). The region of rising solution curves for the ODE, $y' = 1 + t - y$, is the region below the nullcline line, $y = 1 + t$, the region described by $y < 1 + t$.

Problem 3(a).   Problem 3(c).

5. Figure (5), Graph 1 shows the solution curves of the IVP, $y' = -2y + 3e^t$, $y(0) = -5$, $-4, \ldots, 4, 5$, in the rectangle $0 \le t \le 3$, $-5 \le y \le 20$. It appears that the solution curves converge to one another as t increases and eventually curve upwards as if the corresponding solutions are exponentially increasing. The apparent merging of solution curves at the upper right of Fig. (5), Graph 1 is an illusion caused by the finite number of pixels on a computer screen. In Fig. (5), Graph 2 we zoom in on the rectangle $2.996 \le t \le 3.000$, $20.06 \le y \le 20.09$, and we see that the solution curves don't meet. Under the zoom microscope the arcs of the solution curves look like parallel lines. This is no surprise because

over very short ranges for t and for y the rate function $-2y + 3e^t$ doesn't change much, and so the slopes of the curves are nearly identical.

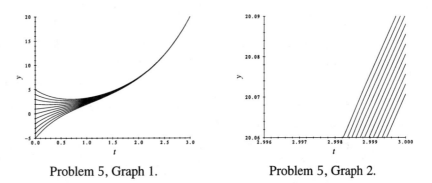

Problem 5, Graph 1. Problem 5, Graph 2.

7. We don't usually say much about group problems, but occasionally we do. If you are looking for a rate function $f(t, y)$ that forces solution curves of $y' = f(t, y)$ to do a lot of twisting, or turning, or explosive growth, try a trigonometric function in t or y, or a polynomial in y. For example, use a numerical solver and take a look at the solution curves of $y' = y^3 \sin t$, or of $y' = 3y \sin y - t \sin t$, in the rectangle $0 \le t \le 10$, $|y| < 5$.

1.3 The Search for Solution Formulas

Comments

Most methods for finding solution formulas for first-order ODEs are based on "preparing" an ODE for application of Antiderivative Theorem 1.3.1. That theorem is really "The Mother of All Solution Methods," even though it is just a slight reformulation of the Fundamental Theorem of Calculus. It is the basis for most techniques for solving first-order ODEs in closed form.

We prepare a first-order linear ODE in normal form for antidifferentiation by multiplying by an integrating factor. The separable ODE, $N(y)y' + M(x) = 0$, is already in shape for antidifferentiation (Section 1.6), as are the exact ODEs of Problem 6 in Section 1.6. The changes of variables in the last section of the chapter are applied to convert an ODE to a form that can then be prepared for antidifferentiation. So, it isn't surprising that antidifferentiation lies at the heart of finding solution formulas for ODEs.

First-order linear ODEs are the most important in applications, and that's why we tackle them first. The integrating factor technique is straightforward, but it takes practice to make the steps of the solution recipe become natural. In this section, the two antiderivatives needed for the technique can be evaluated in terms of familiar functions. Section 1.4 considers some linear ODEs for which the antiderivatives can't be easily calculated. That section takes up other features of linearity that we don't have room to do here.

1. **(a).** The ODE, $y' = \sin t - t^3 y$, is first order (due to the term y'), linear in y, and has normal linear form, $y' + t^3 y = \sin t$.

(c). The ODE, $(t^2 + y^2)^{1/2} = y' + t$, is first order (due to the term y') and nonlinear in y [due to the term $(t^2 + y^2)^{1/2}$].

(e). The ODE, $y'' - t^2 y = \sin t + (y')^2$, is second order (due to y''), and nonlinear in y [due to $(y')^2$].

(g). The ODE, $e^t y'' + (\sin t) y' + 3y = 5e^t$, is second order (due to y''), linear in y, and has normal linear form, $y'' + e^{-t}(\sin t) y' + 3e^{-t} y = 5$.

3.   Replace y by rt^3 in the ODE, and find the values of r for which the equation holds for all t.

     (a). $t^2 y'' + 6ty' + 5y = 0$ becomes $t^2(6rt) + 6t(3rt^2) + 5(rt^3) = 29rt^3 = 0$, which holds for all t if and only if $r = 0$.

     (c). Inserting $y = rt^3$ into the ODE $t^4 y' = y^2$, we have $3rt^6 = r^2 t^6$, and so $3r = r^2$. So, $r = 0, 3$, and $y = 0$, $y = 3t^3$ are solutions.

5.   In (a)–(f) the Antiderivative Theorem 1.3.1 is implemented by antidifferentiating each side of the ODE. C, C_1, C_2, and C_3 are arbitrary constants in the solutions below.

     (a). Antidifferentiating both sides of $y' = 5 + \cos t$ gives $y = 5t + \sin t + C$.

     (c). Antidifferentiating both sides gives $y = \int e^{-t} \cos 2t \, dt$. $\int e^{-t} \cos 2t \, dt$ is of the form $\int e^{at} \cos bt \, dt$. So from Table 1.3.1 we see that the solution is $y = (e^{-t}/5)(-\cos 2t + 2\sin 2t) + C$.

     (e). Antidifferentiating $y'' = \sin t$ twice, we obtain $y = -\sin t + C_1 t + C_2$. Applying the initial conditions, we get the equations $C_2 = 0$ and $C_1 = 2$, so $y = 2t - \sin t$.

     (g). Multiplying the ODE $y'' + y' = e^t$ by e^t, we obtain $(e^t y')' = e^{2t}$. Antidifferentiating, we obtain $e^t y' = e^{2t}/2 + C_1$. So, $y' = e^t/2 + C_1 e^{-t}$. Antidifferentiating again, $y = e^t/2 - C_1 e^{-t} + C_2$.

7.   In each case the equation must first be written in normal linear form, $y' + p(t)y = q(t)$. Then the integrating factor $e^{P(t)}$, where $P(t) = \int p(t) \, dt$, is calculated and each side of the ODE is multiplied by $e^{P(t)}$ to obtain $(e^{P(t)}y)' = e^{P(t)}q$. Antidifferentiating and multiplying by $e^{-P(t)}$ gives the family of solutions $y = e^{-P(t)}C + e^{-P(t)}R(t)$, where $R(t) = \int e^{P(t)}q(t) \, dt$. C is an arbitrary constant, and t lies in an interval for which the ODE is normal and $p(t)$ and $q(t)$ are continuous. It is best to follow the steps of this recipe in each case, rather than memorize the solution formula. The use of integral tables will speed up the calculations. In each case, $-\infty < t < \infty$ unless otherwise noted.

     (a). For $y' - 2ty = t$, the integrating factor is $e^{\int -2t \, dt} = e^{-t^2}$. So, after multiplying by e^{-t^2}, we have that $e^{-t^2}(y' - 2ty) = (e^{-t^2}y)' = te^{-t^2}$. Integrating, we have that $e^{-t^2}y(t) = \int te^{-t^2} \, dt = -e^{-t^2}/2 + C$. The general solution is $y = Ce^{t^2} - 1/2$, where C is any constant and $-\infty < t < \infty$.

     (c). The normal form is $y' + (\sin t)y = \sin t$, so the integrating factor is $e^{\int \sin t \, dt} = e^{-\cos t}$. We obtain $e^{-\cos t}(y' + (\sin t)y) = (e^{-\cos t}y)' = e^{-\cos t} \sin t$. Integrating, $e^{-\cos t}y = e^{-\cos t} + C$ and the general solution is $y = Ce^{\cos t} + 1$ since $\int e^{-\cos t} \sin t \, dt = e^{-\cos t}$.

     (e). The normal form is $y' + ty = t/2$, so the integrating factor is $e^{\int t \, dt} = e^{t^2/2}$. We obtain $e^{t^2/2}(y' + ty) = (e^{t^2/2}y)' = (t/2)e^{t^2/2}$. Integrating, $e^{t^2/2}y = (t/2)e^{t^2/2} + C$ and the general solution is $y = Ce^{-t^2/2} + 1/2$.

9. You may either use graphics to plot the solutions given by the formulas derived in Problem 8, or you may apply a numerical solver directly to the appropriate IVP, as we did. See Figs. 9(a) and 9(c) for the solution curves of the IVPs of Problem 8.

Problem 9(a).   Problem 9(c).

11. First write the ODE in normal form, then find the integrating factor and solve to get the general solution. Use the initial data to evaluate the arbitrary constant C in the general solution formula.

(a). The normal form for $ty' + 2y = \sin t$ is $y' + (2/t)y = (1/t)\sin t$, $t > 0$, and the integrating factor is $e^{\int 2t^{-1}\,dt} = e^{2\ln|t|} = t^2$. We obtain $(t^2 y)' = t\sin t$. Integrating, $t^2 y = \sin t - t\cos t + C$. So $y = (1/t^2)\sin t - (1/t)\cos t + C/t^2$. Since $y(\pi) = 1/\pi$, $1/\pi = 1/\pi + C/\pi^2$, so $C = 0$. The solution is $y = (1/t^2)\sin t - (1/t)\cos t$, $t > 0$. As $t \to +\infty$, $y(t) \to 0$.

(c). The integrating factor for $y' + (\cot t)y = 2\cos t$ is $e^{\int \cot t\,dt} = e^{\ln(\sin t)} = \sin t$. We obtain $((\sin t)y)' = 2\sin t\cos t$. Integrating, $(\sin t)y = \sin^2 t + C$, so $y = \sin t + C\csc t$. Since $y(\pi/2) = 3$, $3 = \sin \pi/2 + C\csc \pi/2 = 1 + C$ so $C = 2$. The solution is $y = \sin t + 2\csc t$, $0 < t < \pi$. As $t \to 0^+$, $y \to +\infty$ since $\csc t \to +\infty$.

13. Group problem.

1.4 Modeling with Linear ODEs

Comments

We like to focus on the total response formula (Theorem 1.4.5) and the Balance Law, because those two items have such widespread applications and interpretations. We have given some of them in this section in the context of mixture problems, and we will include others later in the book.

Other than doing a "good guessing" example, we don't make a big deal about the structure of the solution set of a first order linear ODE because it will come up again in Chapter 3 when we treat second-order linear ODEs, and again in Chapter 7, where linear systems are handled. So why have we put the structure stuff here at all? Mostly for the readers who want to see and understand the theory of a linear ODE, and because it does serve as a neat way to look at the solution formula produced by the recipe of the last section for constructing a general solution.

We always do some models with step function inputs because these inputs are so common in applications, numerical solvers (most of them) can handle these functions, and they are easy to understand and appreciate. As step functions show, a computer can easily handle some things that a solution formula has trouble with.

1. Let's denote the amount of waste (measured in gallons) in the tank at time t (measured in minutes) by $y(t)$. By the Balance Law, we have

$$y' = \text{Rate in} - \text{Rate out}$$

$$= r_{in} - \frac{y(t)}{1000} \cdot r_{out}$$

where r_{in} is the inflow rate of waste (in gal/min), $y/1000$ is the concentration of waste in the tank (in gallons of waste per gallon of mixture), and r_{out} is the outflow rate of contaminated water (in gal/min). For parts **(a)** and **(b)**, $r_{in} = 1 = r_{out}$. At time 0, the water in the tank is clean, so $y(0) = 0$.

(a). The IVP is

$$y' = 1 - \frac{y}{1000} \cdot 1 = 1 - 0.001y, \qquad y(0) = 0$$

In normal linear form the ODE is

$$y' + 0.001y = 1$$

so the integrating factor is $e^{0.001t}$. We obtain $(e^{0.001t}y)' = e^{0.001t}$. Integrating, $e^{0.001t}y = 1000e^{0.001t} + C$, so solutions of the ODE have the form

$$y = Ce^{-0.001t} + 1000$$

Imposing the initial condition $y(0) = 0$, we see that $C = -1000$, so

$$y(t) = 1000(1 - e^{-0.001t}) \text{ gallons}$$

As $t \to +\infty$, $e^{-0.001t} \to 0$, so the pollution becomes total and $y(t) \to 1000$ gallons.

(c). Let's model the sudden change in the waste inflow rate from 1 gal/min to 0 gal/min (the change occurs at $t = 60$ min). So

$$r_{in} = \begin{cases} 1, & 0 \le t \le 60 \\ 0, & t > 60 \end{cases} = \text{step}(60 - t), \quad t \ge 0$$

The modeling IVP then is

$$y' = \text{step}(60 - t) - \frac{y}{1000} \cdot 1$$

$$= \text{step}(60 - t) - 0.001y$$

$$y(0) = 0$$

Applying a numerical solver, we get the graph shown in Fig. 1(c) for the solution $y(t)$.

Since there is no inflow of waste after $t = 60$ min, we expect the amount of waste in the tank to decline exponentially to 0 as $t \to +\infty$. We see from the figure that this exponential decay is pretty slow, and we see from the answer to part **(a)** that this is expected because of the term $e^{-0.001t}$.

Problem 1(c).

3.   Two different ODEs describe the system, before, or after $t = 3$ min. If $S(t)$ denotes the amount of salt in the tank at time t, then for $t < 3$ min, $S' = (2\text{lb/gal})(3\text{gal/min}) = 6$ lb/min. So $S = 6t + C$. Since no salt is present initially, $S(0) = 0 + C$ and $C = 0$. $S = 6t$ is the amount of salt in the tank for $t < 3$ min. Note that at 3 min, $S = 6 \cdot 3 = 18$ lb, which becomes the initial condition for the next time interval ($t \geq 3$).

For $t \geq 3$ min, $S' = (2\text{lb/gal})(3\text{gal/min}) - (S/V)(3\text{gal/min}) = 6 - 3S/V$, where V is the volume of water in the tank. V is constant for $t \geq 3$ since inflow and outflow rates are equal. Using $t_0 = 3$ and an integrating factor $\exp(3t/V)$, $[S\exp(3t/V)]' = 6\exp(3t/V)$, so

$$S = 2V + C\exp(-3t/V), \quad t \geq 3.$$

Since $V = 50 + (3\text{gal/min})(3\text{min}) = 59$ gal and $S(3) = 18$ lbs, we must have $S(3) = 18 = 2(59) + C\exp(-3 \cdot 3/59) = 118 + C\exp(-9/59)$. So, $C = -100\exp(9/59)$ and

$$S = 118 - 100\exp(-3(t - t_0)/59)$$

for $t > 3$.

(a). When $t = 2$ min, $S(2) = 6 \cdot 2 = 12$ lbs salt. When $t = 25$ min, $S = 118 - 100\exp(-3 \cdot 22/59) = 85.3$ lbs salt.

5.   We use the solution formulas found in Problem 4[see "Answers to Selected Problems" at the end of the text for parts **(a)**, **(c)**, and **(e)**] . In every case in this specific problem the general solution y_u of the undriven ODE tends to 0 as $t \to +\infty$ (of course, this will not always be the case). So the general solution of the driven ODE approaches the particular solution found in the corresponding part of Problem 4.

(a). As $t \to +\infty$, all solutions tend to the parabolic curve described by $y_d = t^2 - 2t + 2$.

(c). As $t \to +\infty$, all solutions tend to 0, since $y_d = te^{-2t}$ has this property.

(e). As $t \to +\infty$, all solutions tend to the periodic function of period π, $y_d = -\cos 2t + 2\sin 2t$.

7.   The integrating factor of the undriven linear ODE $y' + 2ty = 0$ is e^{t^2}. So the general solution of that ODE is $y_u = Ce^{-t^2}$. As $t \to +\infty$, $y_u(t) \to 0$. This means that all solutions of the driven ODE, $y' + 2ty = q(t)$, tend to some particular solution $y_d(t)$ since $y(t) = y_u + y_d$. In each case we use a numerical solver for $0 \le t \le 10$ with $y(0) = 0, \pm 2, \pm 4$.

(a). The general solution of $y' + 2ty = 1$ is $y = Ce^{-t^2} + e^{-t^2}\int e^{t^2}\,dt$, but we can't express the integral in terms of any of the elementary functions. So we turn to a numerical solver to solve the IVPs. From Fig 7(a), we conjecture that as $t \to +\infty$, all solutions tend to 0.

(c). The general solution of $y' + 2ty = 1 + t^2$ is $y = Ce^{-t^2} + e^{-t^2}\int e^{t^2}(1 + t^2)dt$. Looking at Fig. 7(c) it looks as if $y(t)$ tends to a straight line as $t \to +\infty$. Measuring the "rise" and the "run" from the figure, we guess that the straight line asymptote has a slope of $1/2$. Moreover, putting a straight-edge on the graph, it looks as if the line would go through the point $t = 0$, $y = 0$. So one might conclude that as $t \to +\infty$ all solutions tend to the line $y = t/2$, although we haven't actually proved this. Note, however, that there is no straight line solution $y(t) = A + Bt$ which solves the ODE.

Problem 7(a).

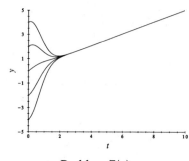

Problem 7(c).

9.   **(a).**   See Fig 9(a) for the graphs of the solutions of the IVPs $y' + y = 0$, $y(-5) = 0, \pm 2, \pm 4$.

(b). See Fig 9(b) for the graph of the solution of $y' + y = t\cos(t^2)$, $y(-5) = 4$.

(c). Fig. 9(a) shows the response $y_h(t) = Ce^{-t}$ to initial data $y(-5) = 0, \pm 2, \pm 4$ with no input, and Fig. 9(b) [see part **(b)**] the total response to initial data $y(-5) = 4$ and to input $q(t) = t\cos(t^2)$. The response to any initial data dies out with increasing time. A sinusoidal response [apparently with rapidly increasing frequency like that of the input $q(t)$] due to the driving term remains. The solution curves in Fig. 9(c) show the superpositions of these two responses. As the initial response dies out, all solutions converge to the driven response.

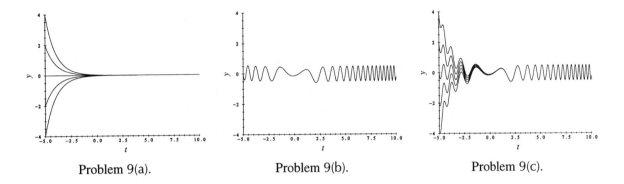

Problem 9(a).   Problem 9(b).   Problem 9(c).

11. Here you can find the solutions of the IVPs explicitly by the integrating factor technique, and then plot the graph. You can also just use a numerical solver/grapher directly without bothering to find explicit solution formulas. We give these formulas below in terms of a constant C, which must be evaluated by using the given initial condition. However, we did the graphs with our numerical solver, ignoring the solution formulas. Your solver may return error messages as ours did in parts **(c)** and **(d)** because of a singularity at the origin [part **(c)**] and at the interval's endpoints [part **(d)**]. In part **(c)** do whatever you can to get the graph of the solution $y = t^2/4$ through the origin.

(a). The integrating factor for $y' = (\sin t)(1 - y)$ is $e^{-\cos t}$, and the general solution is $y = 1 + Ce^{\cos t}$, $-\infty < t < \infty$. See Fig. 11(a) for the solution curves with $y(-\pi/2) = -1$, 0, 1.

(c). The integrating factor for the ODE in normal form, $y' + 2y/t = t$, $t \neq 0$, is t^2, and the general solution is $y = t^2/4 + Ct^{-2}$, $t > 0$, or $t < 0$. See Fig. 11(c) for the solution curves with $y(\pm 2) = 0$, 1, 2. Note that $y = t^2/4$ is the only solution defined for all t. Solutions with positive C go to $+\infty$ as $t \to 0$; solutions with negative C go to $-\infty$.

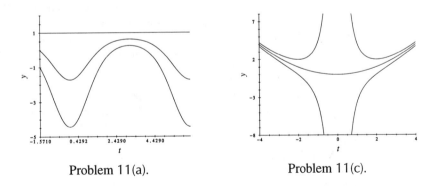

Problem 11(a).   Problem 11(c).

13. The general solution of the linear ODE, $y' + y = q(t)$ is $y = Ce^{-t} + y_d(t)$, where y_d is any particular solution. In this problem we assume that there is some solution $y_d(t)$ that "blows up" as $t \to +\infty$, [i.e., $y_d(t) \to +\infty$ as $t \to +\infty$]. Then, we see from the general solution formula that all solutions have this property since $Ce^{-t} \to 0$ as $t \to +\infty$, and so this term has no effect on the asymptotic behavior as $t \to +\infty$.

15. **(a).** The amount $A(1)$ at the end of one year with 9% continuous compounding is $A(1) = A_0e^{0.09}$ since the model equation is $A' = 0.09A$, $A(0) = A_0$. Setting this equal to the amount after one year with interest payable annually at a rate $r\%$, we have $A_0e^{0.09} = A_0(1 + r/100)$, or $r/100 = e^{0.09} - 1 \approx 0.0942$, (i.e., the annual interest rate would have to be 9.42% to yield the same amount as continuous compounding at 9%).

(c). We must find t such that $2A_0 = A_0e^{kt}$ for $k = 0.05$, 0.09, and 0.12. Canceling A_0, taking logarithms, and solving for t, the respective doubling times are approximately 13.86, 7.70, and 5.78 years.

17. Group project.

1.5 Introduction to Modeling and Systems

Comments

This section is the longest in the text. The main aim of this section is to communicate the idea that one can often create a mathematical model of a natural process if the system is described by a collection of laws or principles. These laws, which are usually based on experiment, observation, and insight, need not be expressed in mathematical terms. The mathematical modeler translates the laws into mathematical equations, ODEs and IVPs, using the methods of this chapter. In this section, the translation process, the solution of the mathematical equations, and the interpretation of the mathematical solution are explained by using the examples of motion along the local vertical and of radioactive decay. The notions of state variables and dynamical systems are introduced (in an imprecise way) so that the reader can see that a finite collection of variables that change in time often is adequate to describe the evolution of a natural process. The exact definition of a dynamical system in terms of a relation such as $y(t_2 + t_1, y_0) = y(t_2, y(t_1, y_0))$, $y(t_0) = y_0$, seems inappropriate at this early stage. In this book dynamical systems are mostly modeled with differential equations. In Chapter 2, when numerical methods for solving ODEs are introduced, discrete dynamical systems are considered (see Section 2.7). Any one-step numerical ODE solver algorithm (e.g., the Euler or Runge-Kutta methods of Section 2.5) can be considered to be a discrete dynamical simulation of the ODE. One may even treat the discrete solver algorithm as a discrete model of the original natural phenomenon.

A secondary aim of this section is to introduce the notion of a differential system in a mild way and to establish the connection between differential systems, modeling, and dynamical systems. The vertical motion model does that in a familiar environment. Section 1.7 takes up systems in a much more serious way.

Modeling is a complex process, and people become adept at it mostly by extensive practice rather than by memorizing modeling rules or schematics. For this reason, it is probably best to work through a specific modeling process rather than to think about modeling principles in the abstract.

An experiment

Throw a whiffle ball up into the air; does it take longer to go up or to come down? Ask a group this question, and we guarantee that the responses will divide four ways: longer going up, longer coming down, equal times, or it all depends. If you are in a room with a high ceiling, throw a real whiffle ball up and ask everyone the same question. You will probably get the same four responses, because it is hard to tell the times by just watching the ball. You may want to devise some way to do alternate experiments with a whiffle ball. Even in this simple setting, the proof that the fall time is longer is hard [Problem 9(**c**)], but the computer simulation (Figures 1.5.3, 1.5.4) shows the answer quite dramatically. See the article, "Projectile Motion with Arbitrary Resistance," Tilak de Alwis, *The College Mathematics Journal*, **26** (1995), 361–367, for a general discussion (with additional references) of vertical motion against resistance.

The other process modeled in the text is radioactive decay and its application to dating. You might find something radioactive and test it with a Geiger counter, but this doesn't sound very appealing. The notion of a half-life is important here. See also the background material for Section 1.1.

---

1. The rate equation is $N' = -kN$, where k is a positive constant. The ODE can be solved by multiplying by the integrating factor e^{kt} and rewriting as $e^{kt} N' + k e^{kt} N = [e^{kt} N]' = 0$. So we have that $e^{kt} N = C$, or $N(t) = Ce^{-kt}$, where C is any constant. The half-life is related to the value of k by $k\tau = \ln 2$. To see this, note that $N(0) = C$. At the time $t = \tau$, $N(0)/2 = N(0)e^{-k\tau}$ so $1/2 = e^{-k\tau}$, or $k\tau = \ln 2$. The data of the problem imply that $0.989 N_0 = N_0 e^{-25k}$, where $N_0 = N(0)$. Canceling N_0, taking logarithms, and solving for the constant k, we have that $k = -(\ln 0.989)/25 = 0.000442$ (years)$^{-1}$ So the half-life is $\tau = (\ln 2)/k \approx 1567$ years.

3. If $N(t)$ is the amount of the phosphorus at time t (in days), then $N(0) = 8$, and $N(t) = 8e^{-kt}$ [from formula (6)]. Now $k = (\ln 2)/14.2 = 0.0488$/day from text formula (7). To find the time t for which $N(t) = 1.0 \times 10^{-5}$, we have $1 \times 10^{-5} = 8e^{-0.0488t}$. Taking logarithms and solving for t, we have $t = 278.5$ days.

5. The model equations are

$$y''(t) = -g, \qquad y(0) = 0, \quad y'(0) = 2000 \text{ cm/sec}$$

where $y(t)$ is the height of the body of mass 600 gm above the ground at time t. The model equations are valid for $0 \le t \le T$, where T is the positive time of impact at which $y(T) = 0$. The value of g is 980 cm/sec^2.

(**a**). Integrating each side of $y'' = -g$ from $t = 0$ to a general value of t, $0 \le t \le T$, we have that $y'(t) - y'(0) = -gt$. Since $v_0 = y'(0) = 2000$ cm/sec and since the velocity $y'(t)$ is 0 at the highest point, $0 - 2000 = -gt$ so $t = 2000/g \approx 2.04$ sec. Integrating $y'' = -g$ twice, we have $y(t) = y_0 + v_0 t - (1/2)gt^2$. With the highest point reached at time $t \approx 2.04$ sec, an initial velocity v_0 of 2000 cm/sec, and a zero initial height y_0, we find that the highest point reached is about 2041 cm.

7. The model IVPs are $s'' = -g$, $s(0) = H$, $s'(0) = 0$ for the position $s(t)$ of the stone (valid for $0 \le t \le T$), and $b'' = -g$, $b(1.5) = H$, $b'(1.5) = -20$ for the position $b(t)$ of the ball (valid for $1.5 \le t \le T$), where $s(t)$ and $b(t)$ are the heights of the stone and the ball above the ground at time t, H is the height of the building, and T is the time the stone and the ball hit the ground. In general, we have $y(t) = y_0 + v_0 t - gt^2/2$ if $y'' = -g$, $y(0) = y_0$, $y'(0) = v_0$, which is the form of both model IVPs.

(a). At impact, $s(T) = H - 4.9T^2$ since the initial velocity of the stone is 0. On the other hand, $b(T) = H - 20(T - 1.5) - 4.9(T - 1.5)^2$ since the initial velocity of the baseball is 20 meters/second and the flight time of the ball is 1.5 seconds shorter than that of the stone. Since both the ball and the stone hit the ground at time T, $s(T) = b(T)$, so $H - 4.9T^2 = H - 20(T - 1.5) - 4.9(T - 1.5)^2$. Canceling H and $-4.9T^2$ and solving for T, we have that $T \approx 3.58$ sec. Now we have that $H - S(T) = 4.9T^2 \approx 4.9(3.58)^2 \approx 62.8$, which is the height of the building in meters.

9.   **(a).** With the origin at ground level and up as the positive direction, the whiffle ball is initially at $y(0) = h$ with initial velocity $y'(0) = v_0$. The whiffle ball undergoes viscous damping due to air resistance, and so we see from (10) that $my'' = -mg - ky'$. Recalling that $y' = v$, we have $y'' = v'$, so $mv' = -mg - kv$. Solving for v', $v' = -g - (k/m)v$. The initial conditions become $y(0) = h$ and $v(0) = v_0$.

(c). The strategy is to find T (the time it takes to reach maximum height) and show that $y(2T) > h$, which implies that $\tau > 2T$, where τ is the time it takes for the whiffle ball to fall back to its initial position $y = h$. Then we will have that the time τ to rise and fall is longer than twice the time T to rise to the maximum height, so the time to fall must be longer than the time to rise.

    Solving the IVP, $v' + v = -g$, $v(0) = v_0$, we obtain $v(t) = -g + (v_0 + g)e^{-t}$. Setting $v(T) = 0$ and solving for T,

$$T = \ln(1 + v_0/g)$$

and so this value of T is the time it takes the whiffle ball to reach maximum height. Now since $y'(t) = v(t) = -g + (v_0 + g)e^{-t}$, and $y(0) = h$. Integrating, we see that

$$y(t) = -gt - (v_0 + g)e^{-t} + v_0 + g + h$$

So evaluating $y(t)$ at $2T$ and substituting in our expression for T, we have

$$y(2T) = -2gT - (v_0 + g)e^{-2T} + v_0 + g + h$$
$$= -2g\ln(1 + v_0/g) - g^2/(v_0 + g) + v_0 + g + h$$

Define the function $f(v_0) = y(2T) - h$, and so

$$f(v_0) = -2g\ln(1 + v_0/g) - g^2/(v_0 + g) + v_0 + g$$

If we can show that $f(v_0) > 0$ for $v_0 > 0$, then $y(2T) > h$. First, we note that $f(0) = -2g\ln 1 - g^2/g + 0 + g = 0$. To show that $f(v_0) > 0$ for $v_0 > 0$ all we need to do is to show that $df/dv_0 > 0$ for $v_0 > 0$:

$$\frac{df}{dv_0} = \frac{-2g}{v_0 + g} + \frac{g^2}{(v_0 + g)^2} + 1$$
$$= \frac{-2g(v_0 + g) + g^2 + (v_0 + g)^2}{(v_0 + g)^2} = \frac{v_0^2}{(v_0 + g)^2} > 0$$

and so $f(v_0)$ is an increasing function of v_0 which is positive if $v_0 > 0$. This shows that $y(2T) > h$ and it takes longer for the whiffle ball to fall through a given distance than it does to rise the same distance. This is reasonable because the gravitational and resistance forces oppose each other during the fall (and so the magnitude of the acceleration is lower), but these two forces act in the same direction during the rise.

11. **(a).** Let $Q(t) = q(t)/q_0$. Then $Q'(t) = q'(t)/q_0$. The IVP $q' = -kq$, $q(0) = q_0$, is changed into the IVP, $Q'(t) = -kQ(t)$, $Q(0) = 1$. Also, a careful reading of Problem 10 reveals that $Q(T) = q_T/q_0 = q'(T)/q'(0) = 13.5/1.69$. Therefore, the IVP in the scaled variable Q is $Q' = -kQ$, $Q(0) = 1$, $T \leq t \leq 0$, and the problem is to find the value of T such that $Q(T) = 13.5/1.69 \approx 7.988$.

13. We have $q(T) = q_0 e^{-kT}$ where T is the time that has passed since the shell was living. Solving for T, we find $T = -(1/k)\ln(q(T)/q(0)) = -(\tau/\ln 2)\ln(q'(T)/q'(0))$, so $T = -5568\ln(0.6 \cdot 13.5/13.5)/\ln 2 = -5568\ln(0.6)/\ln 2 = 4103$ years, which is the approximate age of the shell.

15. This is a hard problem because there doesn't seem to be enough information. And there isn't! The challenge is to assume just enough extra that is plausible and leads to a solution. Different people will make different assumptions and get different times for the start of the snowfall. Here is one set of assumptions and the corresponding solution. Assume that the sidewalk has constant width and the man does not turn back. Let time t in hours be measured forward from 12 noon. Since the snow falls steadily, the depth of the snow at time t in inches is $h(t) = a(t - T)$, where T is the time the snow began to fall [$T < 0$] and a is an unknown rate constant. Let $x(t)$ be the position in blocks of the shoveler at time $t \geq 0$. Then $x(0) = 0$, $x(2) = 2$, $x(4) = 3$. The amount of snow ΔV removed in the time span from t to $t + \Delta t$ as the shoveler moves from x to $x + \Delta x$ is $\Delta V = w \Delta x h(t^*)$, where w is the width of the sidewalk, and t^* is some value of time between t and $t + \Delta t$ [$h(t)$ would be too small and $h(t + \Delta t)$ too large]. Divide ΔV by Δt and let $\Delta t \to 0$ (forcing $t^* \to t$) to obtain the snow removal rate $r = \lim_{\Delta t \to 0} \Delta V / \Delta t = w \lim_{\Delta t \to 0} (\Delta x / \Delta t) h(t^*) = wx'(t)a(t - T)$. Let $b = r/wa$ to obtain the rate equation for the shoveler's walk: $x' = b/(t - T)$. So, $x(t) = b \ln(t - T) + C$, where $t \geq 0$ and C is a constant. Our goal is to find T. We have that $0 = b \ln(-T) + C$, $2 = b \ln(2 - T) + C$, $3 = b \ln(4 - T) + C$. Divide the difference of the first two equations by the difference of the last two to obtain $2 = \ln(\frac{2-T}{-T})/\ln(\frac{4-T}{2-T})$, where we have used $\ln y - \ln z = \ln(y/z)$. Cross-multiply by $\ln \frac{4-T}{2-T}$ and use $2\ln(u) = \ln(u)^2$ for $u > 0$ to obtain $\ln(\frac{4-T}{2-T})^2 = \ln(\frac{2-T}{-T})$. So $(\frac{4-T}{2-T})^2 = \frac{2-T}{-T}$. Clearing fractions, multiplying out and combining terms, we have that $T^2 - 2T - 4 = 0$. So, $T = 1 \pm \sqrt{1+4}$. Since T must be negative, we conclude that the snow began to fall at $T = 1 - \sqrt{5} = -1.236$ hours before noon, in other words, at 10:46 A.M.

Background Material: Modeling Principle

A guiding principle of modeling is Occam's Razor:

> Occam's Razor. What can be accounted for by fewer assumptions is explained in vain by more.

William of Occam (1285–1349) was an English theologian and philosopher who applied the Razor to arguments of every kind. The principle is called the Razor because Occam used it so often and so sharply.

Background Material: Radioactive Decay Law

A primitive form of the decay law would go something like this:

> Radioactive Decay Law. In a sample containing a large number of radioactive nuclei the decrease in the number over time is directly proportional to the elapsed time and to the number of nuclei at the start.

Denoting the number of radioactive nuclei in the sample at time t by $N(t)$, and a time interval by Δt, the law translates to the mathematical equation

$$N(t + \Delta t) - N(t) = -kN(t)\Delta t \tag{i}$$

where k is a positive coefficient of proportionality. Observation suggests that in most decay processes k is independent of t and N. A decay process of this type is said to be of *first order* with *rate constant k*.

The mathematical model (i) of the decay law helps us to spot flaws in the law itself. $N(t)$ and $N(t + \Delta t)$ must be integers, but $k\Delta t$ need not be an integer, or even a fraction. If we want to keep the form of the law, we must transform the real phenomenon into an idealization in which a continuous rather than a discrete amount $N(t)$ undergoes decay. The number of nuclei can be calculated from the mass of the sample at any given time. So there is no need to be specific about the units for N or t at this point. Even if $N(t)$ is continuous, (i) could not hold for arbitrarily large Δt since $N(t + \Delta t) \to 0$ as $\Delta t \to \infty$. Nor does (i) make sense if Δt is so small that no nucleus decays in the time span Δt. The failure of the law for large Δt can be ignored since we are interested only in local behavior in time. The difficulty with small Δt is troublesome. We can only hope that the mathematical procedures we now introduce will lead to a mathematical model from which accurate predictions can be made.

If we divide both sides of formula (i) by Δt and let $\Delta t \to 0$ (ignoring the difficulty with small Δt mentioned above), we have the linear ODE

$$N'(t) = \lim_{\Delta t \to 0} \frac{N(t + \Delta T) - N(t)}{\Delta t} = -kN(t) \tag{ii}$$

Note that k is measured in units of reciprocal time.

Background Material: Radiocarbon Dating

The accuracy of the dating process depends on a knowledge of the exact ratio of radioactive ^{14}C to the carbon in the atmosphere. The ratio is now known to have changed over the years. Volcanic eruptions and industrial smoke dump radioactively dead ^{14}C into the atmosphere and lower the ratio. But the most drastic change in recent times has occurred through the testing of nuclear weapons, which releases radioactive ^{14}C, resulting in an increase of 100% in the ratio in some parts of the Northern Hemisphere. These events change the ratio in the atmosphere and so in living tissue. The variations are now factored into the dating process.

Since the experimental error in determining the rate constants for decay processes with long half-lives is large, we would not expect to use this process to date events in the very recent past. Recent events can be dated if a radioactive substance with a short half-life is involved, e.g., white lead contains a radioactive isotope with a half-life of only 22 years. At the other extreme, radioactive substances such as uranium, with half-lives of billions of years, can be used to date the formation of the earth itself.

1.6   Separable Differential Equations

Comments

People often have difficulty with separable ODEs. If you have trouble separating the ODE into the form $N(y)y' + M(x) = 0$, try to move x and y onto different sides of the equals sign to get $N(y)dy = -M(x)dx$ and then integrate each side. You may still have trouble after getting the ODE separated because you can't find the antiderivatives $G(y)$ of N and $F(x)$ of M. Because the solution is written implicitly as $G(y) + F(x) = C$, it may be hard to find $y(x)$ explicitly (sometimes impossible). These technical difficulties often lead people astray. Consequently, we encourage the use of integral tables, or a CAS (if available), and checking (and rechecking) your algebra.

In this section, we explain the difference between an integral curve [the graph of the level set $G(y) + F(x) = C_0$] and a solution curve [the graph of $y = y(x)$, x in some interval, where $G(y(x)) + F(x) = C_0$]. Because a solution curve is the graph of a differentiable function $y = y(x)$, the curve cannot double back on itself and cannot have vertical tangents.

The modeling/application involves the motion of falling bodies subject to Newtonian damping. Problem 6(d) takes up the "longer to rise or to fall" problem first encountered in Section 1.5, but now in the setting of Newtonian damping. It still takes longer to fall than to rise.

1.   The ODE $y' = 2xy^2$ evidently has the constant solution $y = 0$, for all x. In addition, note that the ODE is separable and can be written in the separable form as $y'y^{-2} = 2x$. For $y \neq 0$, we can integrate the separated ODE, obtaining $-y^{-1} = x^2 + C$, or $y = -(x^2 + C)^{-1}$ where C is any constant.

So the general solution of $y' = 2xy^2$ is given by $y = 0$ and the family of solutions $y = -1/(x^2 + C)$, C any constant. If $C \leq 0$, x must lie in an interval on which $x^2 + C \neq 0$. So for example if $C = -1$, we have three solutions on the three intervals, $x < -1$, $|x| < 1$, and $x > 1$.

3. **(a).** Separating variables in $y' = (y+1)/(x+1)$, we have $(y+1)^{-1}y' - (x+1)^{-1} = 0$, which integrates to $\ln|y+1| - \ln|x+1| = K$. So, $\ln(|y+1|/|x+1|) = K$ and $|y+1| = |x+1|e^K$. That is, $y+1 = \pm e^K(x+1)$, i.e., $y = -1 + C(x+1)$, where we have renamed $\pm e^K$ by C. Since $y(1) = 1$, $1 = -1 + 2C$, so $C = 1$. The solution becomes $y = x$, $-1 < x < \infty$, since y' is undefined at $x = -1$.

   (c). The ODE $y' = ye^{-x}$ separates to $y^{-1}\,dy = e^{-x}\,dx$, which integrates to $\ln|y| = -e^{-x} + C$, so $y = \pm\exp(C - e^{-x})$. The initial data $y(0) = e$ imply that $C = 2$, and we take the positive solution. So, $y = \exp(2 - e^{-x})$ for all x.

   (e). The ODE $y' = -x/y$ separates to $y\,dy + x\,dx = 0$, which integrates to $y^2 + x^2 = C$. Since $y(1) = 2$, $C = 5$ and solving for y the solution is $y = (5 - x^2)^{1/2}$, $|x| \le 5^{1/2}$. See Fig. 3(e) for the solution curve.

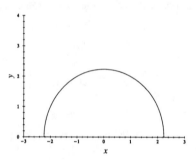

Problem 3(e).

5. **(a).** By formula (15) for the solution of $y' = r(1 - y/K)y$, $y(0) = y_0$, we obtain $y = y_0K/(y_0 + (K - y_0)e^{-rt})$. For the ODE $y' = (1 - y/20)y$, $r = 1$ and $K = 20$. We are also given initial conditions $y_0 = 5, 10, 20$, or 30, so the respective solutions are given by $y = 20y_0/(y_0 + (20 - y_0)e^{-t})$ where $y_0 = 5, 10, 20, 30$, respectively.

   (c). The equilibrium (i.e., constant) solutions of $y' = 3(1 - y/12)y - 8$ are found by setting y' equal to zero and using the quadratic formula (or factoring) to find the roots. We have $3(1 - y/12)y - 8 = -\frac{1}{4}(y^2 - 12y + 32) = -\frac{1}{4}(y - 4)(y - 8)$, and so the equilibrium levels are $y = 4, 8$. This means that if y_0 lies between 0 and 4, the species falls away from the equilibrium $y = 4$ and dies out since $y' < 0$ in this region (e.g., $y_0 = 2$), while if y_0 is between 4 and 8 (e.g., $y_0 = 6$) the population follows a logistic path and is asymptotic upward to the equilibrium level $y = 8$ since $y' > 0$ in this region. If y_0 exceeds 8 (e.g., $y_0 = 10$), the population falls asymptotically toward the equilibrium level $y = 8$ since $y' < 0$ in this region.

   (e). The logistic model is $y' = r(1 - y/K)y$, where $K = 5 \times 10^8$ and $r = 0.01$/day. The solution is given by formula (15) in the text: $y(t) = y_0K/(y_0 + (K - y_0)e^{-0.01t})$. Set $t = 2$, $y_0 = 10^8$ and obtain $y(2) \approx 1.016 \times 10^8$ individuals.

7. Observe that the ODE $v' = -g + (k/m)v^2$ is separable. Separating the variables, we obtain

$$\int \frac{1}{-g + (k/m)v^2} \, dv = \int dt$$

Using partial fractions to carry out the left-hand integration, we have

$$\frac{1}{2} \left(\frac{m}{gk} \right)^{1/2} \ln \left| \frac{v - (mg/k)^{1/2}}{v + (mg/k)^{1/2}} \right| = t + C$$

Plugging in the initial condition $v(0) = 0$ gives $C = 0$. Setting $A = 2(gk/m)^{1/2}$ and taking exponentials of both sides we obtain

$$\left| \frac{v - (mg/k)^{1/2}}{v + (mg/k)^{1/2}} \right| = e^{At}$$

Note that the quantity between the absolute value bars is negative for all $t \geq 0$, and so changing signs and dropping the absolute value, we solve for v to obtain

$$v(t) = \left(\frac{mg}{k} \right)^{1/2} \frac{e^{-At} - 1}{e^{-At} + 1}$$

9. **(a).** The rate function for the ODE $y' = y(1 - y/10) - 9/10$ is the quadratic, $-y^2/10 + y - 9/10$, which factors to yield $y' = -(y-1)(y-9)/10$, $y(0) = y_0$. The equilibrium populations are $y = 1$ and $y = 9$ because y' is zero if $y = 1, 9$. From the factorization of the rate function in the IVP we see that the population declines if $y > 9$ or if $0 < y < 1$ because y' is negative in those regions; $y(t)$ increases if $1 < y < 9$. Separating variables in the ODE and using partial fractions,

$$\frac{1}{y-1} \cdot \frac{1}{y-9} \, dy = -\frac{1}{10} \, dt, \quad y \neq 1, 9$$

$$\frac{1}{8} \left(\frac{1}{y-9} - \frac{1}{y-1} \right) dy = -\frac{1}{10} \, dt$$

$$\ln \left| \frac{y-9}{y-1} \right|^{1/8} = -\frac{1}{10} t + b, \quad b \text{ a constant}$$

$$\left| \frac{y-9}{y-1} \right|^{1/8} = e^b e^{-t/10}$$

$$\frac{y-9}{y-1} = Be^{-4t/5}, \quad B = \pm e^{8b}$$

Solving for y and substituting initial conditions $y(0) = y_0$, we obtain

$$y = \frac{9 - Be^{-4t/5}}{1 - Be^{-4t/5}}, \quad B = \frac{y_0 - 9}{y_0 - 1}, \quad y_0 \neq 1$$

11. Group project.

1.7   Planar Systems and First-Order ODEs

Comments

Planar systems come up in this section in a completely natural way—as a reformulation of second-order ODEs that were treated in Section 1.5. Later we show how to use planar systems to visualize solution curves of first-order ODEs; a big advantage of this approach is that one can immediately "see" the maximally-extended solution curves of the first-order ODE. Another advantage of this approach is that a contour plotter need not be used even when an integral of the first-order ODE is known. This neat trick is very useful in many applications, but the concept may seem rather tricky at first.

Another connection between second-order ODEs and first-order ODEs is made when we show how to reduce the second-order ODEs $y'' = F(t, y')$ and $y'' = F(y, y')$ to systems of first-order ODEs that can be solved one at a time. The algebraic complications in actually constructing the solutions can be difficult. If you use a computer solver to plot solution curves of a second-order ODE, you may also want to use the solver for the corresponding system and then plot ty-component graphs.

For the second-order ODEs in y solved here, solution curves in the ty-plane can cross each other. This is because initial position *and* initial velocity are needed to specify a unique solution of a second-order ODE. If distinct solution curves cross, they do so with different slopes.

The section has a proof of the existence of the escape velocity from a massive body. Besides being a nice example of the use of the reduction of order technique, it also provides a striking example of what the computer cannot do i.e., prove the existence of an escape velocity. The section ends with a run-through of a Lanchester combat model, which starts with a system and ends with a separable ODE. Some people are put off by the use of mathematics in this setting. Finally, take a look at the background material on Einstein, general relativity, and the big bang.

---

1.   Our basic strategy is to find integral curves of $N(x, y)y' + M(x, y) = 0$ by finding solutions of the system $dx/dt = N(x, y)$, $dy/dt = -M(x, y)$. The system can be used to generate plots on a computer solver. $\int N(y)\, dy + \int M(x)\, dx$ is an integral of the ODE when $N(x, y) = N(y)$ and $M(x, y) = M(x)$.

(a). $y - 1/3y^3 + 1/3x^3 - x$ is an integral of the separable ODE $(1 - y^2)y' + x^2 - 1 = 0$. Plots of some integral curves are shown in Fig. 1(a).

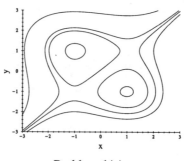

Problem 1(a).

3. **(a).** From Fig. 3(a), we see that y loses if $x(0) = 10$ and $y(0) = 7$ since $y(t) \to 0$ as t increases, while $x(t) \to 4.4$.

Problem 3(a).

5. **(a).** Let $y' = v$, $y'' = v\,dv/dy$. The ODE $2yy'' + (y')^2 = 0$ becomes $2yv\,dv/dy + v^2 = 0$. So $v = 0$ for all t, or $2y\,dv/dy + v = 0$. If $v = 0$ for all t, then y is a constant, but the initial condition $y'(0) = -1$ implies that y is nonconstant. So, v is not always 0. Separating variables, we have $2v^{-1}dv + y^{-1}dy = 0$. Integrating, $\ln|v^2 y| = K$, so $v^2 y = \pm e^K$. So, $v = y' = K_1 y^{-1/2}$. Separating variables, $y^{1/2}dy = K_1\,dt$. Integrating, $2y^{3/2}/3 = K_1 t + K_2$ or $y = (C_1 t + C_2)^{2/3}$, where C_1 and C_2 are constants to be determined by the initial data $y(0) = 1$, $y'(0) = -1$. So, $C_2 = 1$, $C_1 = -3/2$. The solution is $y = (-3t/2 + 1)^{2/3}$.

   (c). Let $y' = v$, $y'' = v'$. The ODE $y'' = -g - y'$ becomes $v' = -g - v$ or $(g + v)^{-1}dv = -dt$. Integrating, $\ln|g + v| = -t + K$, so $g + v = C_1 e^{-t}$. Since $v = y'$, we have $y' = C_1 e^{-t} - g$. Integrating, $y = -C_1 e^{-t} - gt + C_2$. Since $y = h$ and $y' = 0$, at $t = 0$, we have $y = -ge^{-t} - gt + h + g$.

7. The normalized ODE is $y'' + 4t^{-1}y' + 2t^{-2}y = t^{-2}\sin t$, $t > 0$. Following the technique described in Problem 6 and using the fact that $z = t^{-2}$ is a solution of $z'' + 4t^{-1}z' + 2t^{-2}z = 0$ (check by direct substitution), we see that $y = t^{-2}u(t)$, where $u(t)$ solves $t^{-2}u'' + (-4t^{-3} + 4t^{-3})u' = t^{-2}\sin t$, solves the normalized ODE. This last equation simplifies to $u'' = \sin t$. After two integrations, we have $u = -\sin t + C_1 t + C_2$, where C_1 and C_2 are any constants. So $y = -t^{-2}\sin t + C_1 t^{-1} + C_2 t^{-2}$, $t > 0$, gives a family of solutions of the ODE.

9. **(a).** The mass of the object times its acceleration equals the net force acting on the object. The equation of motion is $mz'' = -mM\tilde{G}/(z + R)^3$. Let $v = z'$, $v\,dv/dz = z''$ and separate the z and v variables to obtain $v\,dv = -M\tilde{G}(z + R)^{-3}dz$, which integrates to $v^2 = M\tilde{G}(z + R)^{-2} + v_0^2 - M\tilde{G}R^{-2}$ where it is assumed that at $t = 0$, $v = v_0$ and $z = 0$. So, v remains positive for all z if $v_0 > (M\tilde{G})^{1/2}/R$, which is the escape velocity.

11. Group project.

Background Material: Einstein's Field Equations of General Relativity

These equations are a complicated system of nonlinear partial differential equations. In 1922, the Soviet mathematician Alexander Alexandrovich Friedmann succeeded in obtaining cosmological solutions governing the behavior of the universe as a whole. He assumed that on a broad scale the universe is homogeneous and isotropic and that pressure can be ignored (a dubious hypothesis if the universe is "small"). The PDEs reduce to a single nonlinear IVP

$$2RR'' + R'^2 + kc^2 = 0, \qquad R(t_0) = R_0 > 0, \quad R'(t_0) = v_0 > 0 \qquad \text{(i)}$$

where R is the "radius of the universe," $k = +1$ (spherical geometry), 0 (Euclidean geometry), or -1 (pseudospherical geometry), c is the speed of light, and the derivatives are with respect to time.

For each of the three cases, $k = 0, +1, -1$, decide whether the universe is born with a "big bang" [i.e., at some time t_1, as $t \to t_1^+$, $R(t) \to 0^+$, and $R'(t) \to +\infty$]. In each case, what happens to the universe as t increases? Does the universe die in a "big crunch" [i.e., for some t_2, as $t \to t_2^-$, $R(t) \to 0^+$, and $R'(t) \to -\infty$]. In carrying out this project, address the following points.

- Multiply the ODE in (i) by R' and rewrite as $[R(R')^2 + kc^2 R]' = 0$. Show that $R(R')^2 + kc^2 R = R_0 v_0^2 + kc^2 R_0 = C_0$.

- Set $k = 0$ (a Euclidean universe) and show that $R^{3/2} = R_0^{3/2} \pm 1.5 C_0^{1/2}(t - t_0)$. Show that the minus sign leads to the eventual collapse of the universe (the "big crunch") at some time $T > t_0$, while the plus sign leads to perpetual expansion. What happens in both cases as $R_0 \to 0^+$?

- Set $k = 1$ (a spherical universe). Solve the ODE, $R(R')^2 + c^2 R = C_0$, in terms of a parameter u, by setting $R = C_0 c^{-2} \sin^2 u$, rewriting the ODE with u as the dependent variable, and solving to find t in terms of u. Explain why $t = t(u)$, $R = R(u)$ are the parametric equations of a cycloid in the tR-plane. Interpret each arch of the cycloid in terms of a big bang and big crunch cosmology.

- Set $k = -1$ (a pseudospherical universe). Follow the steps of the spherical case, but set $R = C_0 c^{-2} \sinh^2 u$. Interpret the solution in cosmological terms.

- Rewrite the original ODE as a system of ODEs in dimensionless variables: $dx/ds = 2y$, $dy/ds = -x^{-1}(k + y^2)$, where $x(s_0) = 1$, $y(s_0) = v_0/c$, $x = R/R_0$, $y = R'/c$, $s = ct/2R_0$. Now use an ODE solver/graphics package to plot solutions $x = x(s)$ and $y = y(s)$ as functions of s for $k = 0, +1, -1$. Interpret the graphs in cosmological terms. Discuss the advantage of scaling the original variables as indicated before computing.

1.8 Cold Pills

Comments

Mathematically speaking, the solution techniques of this section are all based on the integrating factor method for a first-order linear ODE (see Section 1.4). These are now applied to a cascade of two or more first-order linear ODEs, where the ODEs can be solved from the top down, one ODE at a time. Subsequent ODEs in the cascade make use of the solutions of some of the preceding ODEs. This process shows that these systems of ODEs are not at all fearsome, even though the algebra can be a little bit complicated.

On a modeling level, the first-order systems of ODEs that represent the flow of medications through the compartments of the body seem quite natural to many people, who find these systems easier to understand than the mechanical, gravitational, or electrical systems traditionally used in ODE courses. We like to introduce systems of ODEs naturally and early in the course, but informally. This approach accustoms everyone to seeing, using, and solving systems, but without all the vocabulary and theory of the full treatment of a system of n ODEs (which we do in Chapter 5). We talk about systems in other sections of Chapter 1, but this is the only section of the chapter devoted entirely to phenomena that can only be modeled using systems.

We give informal definitions of a general compartment system and a linear cascade compartment system, and note how they can be represented by labeled boxes and arrows. By careful examination of the assemblage, a system of rate equations can be written for the flow of the substance being tracked through the compartments. There is a different interpretation of a compartmental system where there is just one physical compartment containing a substance which decays over time into other substances [e.g., radioactive decay (see Problem 14 in Section 1.5 or chemical reactions (see Chapter 5)].

The sensitivity of a system to changes in a rate coefficient is shown in the cold pill models. The clearance coefficient for removing medication from the blood may be much lower for older people than for the young, and we discuss the consequences (see also Problems 5–7). The model that represents taking a fast-dissolving cold pill every four or six hours leads naturally to a system of linear ODEs with a periodic pulse input function of the form A sqw(t,d,T). Functions like this appear throughout the applications and we introduce them here in a natural setting. The system is most easily solved by using a numerical solver with piecewise continuous functions in its predefined list.

---

1. All the algebraic steps are given for part **(a)**.

(a). Analysis of the cascade model as in Example 1.8.8 produces the linear system $x' = -3x$, $y' = x$, $z' = 2x$. The system is solved from the top down. Using the techniques of Section 1.4, we see that the general solution to the first ODE is $x = Ce^{-3t}$. Since $x(0) = 1$, $C = 1$ and the solution to the IVP is $x = e^{-3t}$. Substituting this into the other equations of the system, we have $y' = e^{-3t}$ and $z' = 2e^{-3t}$. Integrating each of these gives $y = -e^{-3t}/3 + C_1$ and $z = -2e^{-3t}/3 + C_2$. Using the initial data, $y(0) = 0 = -1/3 + C_1$, so $C_1 = 1/3$; and $z(0) = 0 = -2/3 + C_2$, so $C_2 = 2/3$. The solutions are $x = e^{-3t}$, $y = (1 - e^{-3t})/3$, and $z = 2(1 - e^{-3t})/3$. As $t \to +\infty$, $x \to 0$, $y \to 1/3$, and $z \to 2/3$.

(c). The system, now with a sinusoidal input, is $x' = 1 + \sin t - 3x$, $y' = 3x - y$. We will solve from the top down. Using the integrating factor e^{3t}, $(xe^{3t})' = (1 + \sin t)e^{3t}$. Integrating, $x = 1/3 + (3 \sin t - \cos t)/10 + Ce^{-3t}$. Using $x(0) = 0$, we get $C = -7/30$ and $x = 1/3 + (3 \sin t - \cos t)/10 - 7e^{-3t}/30$. Since $y' = 3x - y$, another use of the integrating factor method (after replacing x by the expression in t just derived) gives $y = 1 + 3(\sin t - 2 \cos t)/10 + 7e^{-3t}/20 + Ce^{-t}$. So, since $y(0) = 0$, $y = 1 + 3(\sin t - 2 \cos t)/10 + 7e^{-3t}/20 - 3e^{-t}/4$. As $t \to +\infty$, x oscillates about $1/3$ and y oscillates about 1.

3. **(a).** Using integrating factors, solve the first ODE of system (3), and then the other.

The IVP, $dx/dt = -k_1 x$, $x(0) = A$, gives $x = Ae^{-k_1 t}$. Substituting this into the ODE, $dy/dt = k_1 x - k_2 y$, and rearranging, $y' + k_2 y = Ak_1 e^{-k_1 t}$. Using the integrating factor $e^{k_2 t}$, $(ye^{k_2 t})' = Ak_1 e^{(k_2 - k_1)t}$, giving $y = Ak_1 e^{-k_1 t}/(k_2 - k_1) + Ce^{-k_2 t}$, where C is constant. Using $y(0) = B$, we get $B = Ak_1/(k_2 - k_1) + C$, so $C = B - Ak_1/(k_2 - k_1)$. So, $x = Ae^{-k_1 t}$ and $y = Ak_1(e^{-k_1 t} - e^{-k_2 t})/(k_2 - k_1) + Be^{-k_2 t}$.

5.   **(a).** See Fig. 5(a) for antihistamine levels (the bottom curve corresponds to the smallest value of k_1, 0.0691, and the top curve to the largest, 1.5). The curves for the larger values of k_1 eventually cross the curves for the smaller values. This models the fact that effective clearance of the antihistamine from the blood occurs only after the medication level in the blood has peaked; if k_1 is small, it takes longer to peak.

Problem 5(a).

7.   **(a).** The model IVP for the amounts $x(t)$ and $y(t)$ of continuous-acting medication in the GI tract and bloodstream, respectively, is $x'(t) = I - k_1 x$, $x(0) = 0$, and $y'(t) = k_1 x - k_2 y$, $y(0) = 0$.

(c). As $t \to \infty$, $x(t) \to I/k_1$ and $y(t) \to I/k_2$.

(e). See Fig. 7(e), Graph 1 for decongestant levels and Fig. 7(e), Graph 2 for antihistamine levels. As expected, the levels of both medications in the blood continue to rise, particularly the antihistamine.

Problem 7(e), Graph 1.   Problem 7(e), Graph 2.

1.9 Change of Variables and Pursuit Models

Comments

A widely used method to deal with a difficult equation is to change the variables, hoping that the transformed equation will be simpler (and more easily solvable) in the new variables. There are hundreds of changes of variables for dealing with complicated (often nonlinear) ODEs; see, for example, *Handbook of Exact Solutions for Ordinary Differential Equations*, A.D. Polyanin, V.F. Zaitsev; CRC Press Inc., 1995, Boca Raton. See also D. Zwillinger's *Handbook of Differential Equations*, 2nd ed., 1992; Academic Press, San Diego. In this section, we consider a few variable changes that reduce some simple (but nonlinear) first-order ODEs to forms that can be solved by methods of this chapter. Bernoulli ODEs (Problem 9), Riccati ODEs (Problem 10), ODEs involving a function that is homogeneous of order 0 (Example 1.9.2)—these are all transformed to simpler ODEs by a variable change. Applications of variable changes are made to rescaling a model of vertical motion (Example 1.9.5, and Problem 8).

An interesting sidelight to the pursuit problem: from Example 1.9.4 and Problem 11 we see that when the goose reaches the nest it is flying directly into the wind. Is this always true?

As in Section 1.6, it often happens that solutions can only be defined by implicit formulas. So we get integral curves as graphs, rather than solution curves. A solution curve is an arc of an integral curve with no vertical tangents.

---

1. In the solutions below, each ODE is written in the form $y' = f(x, y)$, and it is shown that f is homogeneous of order 0 by verifying that $f(kx, ky) = f(x, y)$. The change of variable $y = xz$ converts the ODE into a separable ODE in x and z, and this new ODE is given, along with its solutions. Finally, the solutions of the original ODE are given, sometimes implicitly, by replacing z by y/x. **Caution:** There is a lot of algebraic manipulation here, so watch your steps!

(a). The ODE is $y' = (y + x)/x$. The rate function is homogeneous of order 0, because

$$f(kx, ky) = \frac{kx + ky}{kx} = \frac{x + y}{x} = f(x, y)$$

If $y = xz$, then $y' = xz' + z = (xz + x)/x = z + 1$, and $xz' = 1$. The solutions are $z = \ln|x| + C$. Using $y = xz$, $y = x\ln|x| + Cx$, where the domain is restricted to $x > 0$ or $x < 0$.

(c). The ODE may be written as $y' = (x^2 - xy - y^2)/(xy)$, and we have that

$$f(kx, ky) = \frac{k^2x^2 - k^2xy - k^2y^2}{k^2xy} = f(x, y)$$

So the rate function is homogeneous of order 0. The variable change $y = xz$ leads to the ODE $xz' + z = (x^2 - x^2z - x^2z^2)/x^2z = (1 - z - z^2)/z$, or $z'z/(1 - z - 2z^2) = 1/x$. This ODE integrates (after using partial fractions or integral tables) to $-\frac{1}{6}\ln|1 - 2z| - \frac{1}{3}\ln|1 + z| = \ln|x| + C$. Let $z = y/x$, rearrange, exponentiate, and obtain $x^3(x - 2y)(x + y)^2 = C$, where C is a constant.

(e). The ODE is $y' = 4 + 7y/x + 2y^2/x^2$, and we have that f is homogeneous of order 0 because

$$f(kx, ky) = 4 + \frac{7ky}{kx} + \frac{2k^2 y^2}{k^2 x^2} = f(x, y)$$

Let $y = xz$; $xz' + z = 4 + 7z + 2z^2$ and $dz/(4 + 6z + 2z^2) = dx/x$, which integrates to $[\ln |(z+1)/(z+2)|]/2 = \ln |x| + K$, which in terms of x and y reduces to $\ln |(x+y)/(2x+y)|^{1/2} - \ln |x| = K$ or $\ln |(x + y)/(2x + y) \cdot (1/x^2)|^{1/2} = K$. Exponentiating, $x + y = Cx^2(2x + y)$, where C is any constant, or $y = (2Cx^3 - x)/(1 - Cx^2)$, where x is such that $1 - Cx^2 \neq 0$.

3.   In each case we find a "general" implicit solution in the form $H(x, y) = C$. We don't always attempt to solve for y in terms of x and C.

(a). Let $z = x + y$, in the ODE, $dy/dx = \cos(x + y)$. Then $dz/dx = 1 + dy/dx = 1 + \cos z$, so $(1 + \cos z)^{-1} dz = dx$. Using a table of integrals, we have $\tan(z/2) = x + C$, so $\tan[(x + y)/2] = x + C$. Then $x + y = 2 \arctan(x + C) + n\pi$, so $y = -x + 2 \arctan(x + C) + n\pi$, where C is any constant and n is any integer.

(c). Let $z = x + 2y$, so $dz/dx = 1 + 2dy/dx$. Then the ODE, $(x + 2y - 1) dx + 3(x + 2y) dy = 0$, becomes $(z - 1) + 3z(dz/dx - 1)/2 = 0$ or $3z \, dz/(z + 2) = dx$, which integrates to $3(z - 2 \ln |z + 2|) = x + C$. In terms of x and y, $3(x + 2y - 2 \ln |x + 2y + 2|) = x + C$, where C is any constant and $|x + 2y + 2| \neq 0$.

5.   **(a).** Make the change of variable $z = 1/y$ in the logistic equation $y' = r(1 - y/K)y$. Since $y = 1/z$, we see that $y' = -z^{-2}z'$, so

$$-z^{-2}z' = r\left(1 - \frac{1}{Kz}\right)\left(\frac{1}{z}\right)$$

$$-z' = r\left(z - \frac{1}{K}\right)$$

$$z' = -rz + \frac{r}{K}$$

which is a linear ODE.

7.   Set $x(t) = P(t)/K$ and $t = s/r$ in the logistic ODE $P'(t) = r(1 - P(t)/k)P(t)$. By the chain rule,

$$\frac{dx}{ds} = \frac{1}{K}\frac{dP}{dt}\frac{dt}{ds} = \frac{1}{K}r(1 - p/K)p/r$$

$$= \frac{1}{K}(1 - x)Kx = (1 - x)x$$

The advantage of using the scaled ODE is that there are no parameters in the rate function. This means that a solution of the scaled equation can be used to describe the behavior of solutions of the unscaled ODE regardless of the values of the positive constants r and K.

9.   The Bernoulli ODE is $dy/dt + p(t)y = q(t)y^b$.

(a). Let $b \neq 0, 1$ and divide the ODE $dz/dt + (1 - b)p(t)z = (1 - b)q(t)$ through by y^b.

Then multiply by $1 - b$ and the ODE can be written as

$$\frac{d}{dt}(y^{1-b}) + (1 - b)p(t)y^{1-b} = (1 - b)q(t)$$

Now make the change of variables $z = y^{1-b}$ and the ODE takes the form

$$\frac{dz}{dt} + (1 - b)p(t)z = (1 - b)q(t) \tag{i}$$

which is a linear ODE. If $y(t)$ solves the original ODE, then $z(t) = (y(t))^{1-b}$ solves the linear ODE (i). Conversely, if z is any solution to the last ODE, then $y = z^{1/(1-b)}$ is a solution of the original ODE, or perhaps several solutions after taking the root $1/(1 - b)$.

(c). We may transform the Bernoulli ODE, $y' + t^{-1}y = y^{-4}$, to a linear ODE using the variable change $z = y^5$. We obtain $z' + 5t^{-1}z = 5$, where an integrating factor is $e^{5\ln t} = t^5$, and the solutions are given by $z = 5t/6 + C/t^5$. Then $y = z^{1/5} = (5t/6 + C/t^5)^{1/5}$, where $t > 0$ and C is an arbitrary constant. See Fig. 9(c) for solution curves.

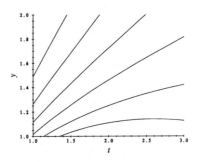

Problem 9(c).

11. **(a).** The problem may be set up exactly as the pursuit problem in the text, except that the wind's velocity has component $-b/2$ along the x-axis and $b/2$ along the y-axis. So, $x' = -b\cos\theta - b/2$ and $y' = -b\sin\theta + b/2$. Dividing, we have

$$\frac{y'}{x'} = \frac{dy}{dx} = \frac{2\sin\theta - 1}{2\cos\theta + 1} = \frac{2y - (x^2 + y^2)^{1/2}}{2x + (x^2 + y^2)^{1/2}}$$

Since this quotient is a homogeneous function of order 0 in x and y, we may introduce $y = xz$ to obtain the separable ODE $x\,dz/dx + z = [2z - (1 + z^2)^{1/2}]/[2 + (1 + z^2)^{1/2}]$, or

$$\frac{2 - (1 + z^2)^{1/2}}{(z + 1)(1 + z^2)^{1/2}}\,dz = \frac{1}{x}\,dx$$

Let $z = u - 1$ to reduce the left side to a form found in integral tables. After solving and returning to x, y variables and simplifying, we obtain $[\sqrt{2}(x^2 + y^2)^{1/2} + x - y]^2 = a^{\sqrt{2}}(3 + 2\sqrt{2})|x + y|^{2-\sqrt{2}}$, where the initial data are used. The goose follows the path defined by this expression if it heads towards its nest while the wind blows from the southeast.

Background Material: ODEs in Polar Coordinates

The ODE in rectangular coordinates x and y

$$dy/dx = f(x, y) \tag{i}$$

can be transformed into an equivalent ODE in polar coordinates r and θ. First recall the relations between polar and rectangular coordinates:

$$x = r\cos\theta, \qquad y = r\sin\theta, \qquad r = (x^2 + y^2)^{1/2}, \qquad \tan\theta = y/x$$

Now suppose that $y = g(x)$ defines a solution curve of the ODE (i). This solution curve can also be defined (perhaps only in part) by specifying r as a function of the polar angle θ, that is, $r = r(\theta)$. In polar coordinates the relation $y = g(x)$ becomes

$$r\sin\theta = g(r\cos\theta) \tag{ii}$$

Differentiating each side with respect to θ and using the Chain Rule, we have

$$\frac{dr}{d\theta}\sin\theta + r\cos\theta = g'(r\cos\theta)\left(\frac{dr}{d\theta}\cos\theta - r\sin\theta\right)$$

Since $y = g(x)$ solves $y' = f(x, y)$, then $g'(r\cos\theta) = f(r\cos\theta, r\sin\theta)$. So

$$\frac{dr}{d\theta}\sin\theta + r\cos\theta = f(r\cos\theta, r\sin\theta)\left(\frac{dr}{d\theta}\cos\theta - r\sin\theta\right) \tag{iii}$$

The ODE in (iii) is equivalent to $y' = f(x, y)$. Although ODE (iii) looks complicated, it is actually simpler than (i) for some functions $f(x, y)$. Let's look at an example.

Applying formula (iii) to the ODE

$$\frac{dy}{dx} = \frac{y - 2x}{2y + x} \tag{iv}$$

we have the equivalent ODE in polar coordinates:

$$\frac{dr}{d\theta}\sin\theta + r\cos\theta = \frac{r\sin\theta - 2r\cos\theta}{2r\sin\theta + r\cos\theta}\left(\frac{dr}{d\theta}\cos\theta - r\sin\theta\right) \tag{v}$$

After canceling the common factor r in the quotient, multiplying each side by $2\sin\theta + \cos\theta$, simplifying, and again canceling a common factor, ODE (v) reduces, after some algebra, to the linear ODE in r:

$$\frac{dr}{d\theta} = -\frac{r}{2} \tag{vi}$$

The solutions of ODE (vi) are given by

$$r = Ce^{-\theta/2}$$

where C is any nonnegative constant.

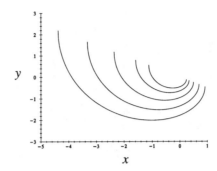

The corresponding curves in the xy-plane are spirals around the origin. The five solution curves shown are arcs of one of these spirals, arcs that have no vertical tangent line because only such arcs can be graphs of functions of x. The "top" arcs of the spiral are also solution curves, but they are not shown. Finally, note that taking logarithms of each side of the equation $r = Ce^{-\theta/2}$ and returning to the xy-variables, the solutions $y(x)$ of ODE (iv) satisfy the equation

$$\frac{1}{2}\ln(x^2 + y^2) = \ln C - \frac{1}{2}\arctan(y/x) \tag{vii}$$

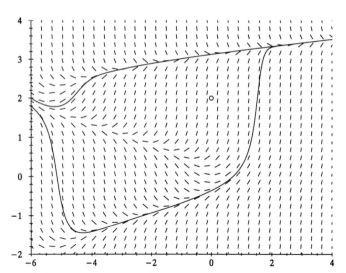

Initial Value Problems and Their Approximate Solutions

The goal of this chapter is to deepen and extend some topics that were introduced in Chapter 1: existence and uniqueness of a solution for a first-order IVP, numerical approximation of a solution of an IVP, and a study of sensitivity via computer simulation. The early sections of the chapter have a bare-bones treatment of the topics, with many examples, and this should suffice for the most part. The last section of the chapter has a different character from the rest of the chapter in that the object of prime interest is a discrete dynamical system. Euler's Method applied to the logistic ODE is basically only a convenient artifice for initiating this study. We think that this section is a lot of fun, bringing in period doubling and apparently chaotic wandering.

We have tried to be mindful throughout that although numerical methods are important because they drive the engine that produces graphs of solutions, they are not the main object of interest in the course. Therefore we have tried to give numerical methods a featured spot as early as possible in the text, but we limit our coverage to concepts that would affect the ability to produce decent graphs. There are many numerical solvers available on the market today which can generate most of the graphs of solutions of ODEs which appear in this text. Most CAS packages contain numerical ODE solvers as well, but because of their symbolic manipulation facilities they are also capable of producing solution formulas for some classes of ODEs. A brief overview article on ODE solvers appeared in 1994 in *The College Mathematics Journal*'s Special Issue on Differential Equations.[1] Along with a short description of each solver, the article contains information about the academic cost of the package and a telephone number for obtaining more information (as of 1994).

The graphs your solver produces may look different from the graphs in the text. There are many reasons for this: sometimes the reasons are platform dependent, sometimes algorithm dependent,

---

[1] Andrew Flint and Ron Wood, "ODE Solvers in the Classroom", *College Math. Jour.*, *25*, No. 5 (1994), pp. 458–461.

and sometimes maybe you (or we) just hit the wrong keys on the keyboard. The way your machine does arithmetic and function evaluations is an example of the former. Memory may also be a problem (the data sets for some of our graphs are huge). If you have an adaptive solver (as we do), then your default settings may cause you to miss some important features in the graph. In any case, there are some things you can do to improve the appearance of your graphs. For Runge-Kutta solvers try changing the step size and shortening the solve-time interval. For adaptive solvers try changing the default settings for the absolute and relative error bounds, the number of plotted points, the maximum number of function evaluations, the maximum and minimum interval step size, etc. Please keep in mind that the graphs that appear in this text are not the result of a first pass. Often, we had to adjust the time span, the intervals for the state variables, the initial data, or the solver parameters before we could get a picture that was both informative and good-looking. Another reason why your graphs will often not look as good as those in the text: the printing technology that produced the graphs in the text is equivalent to a 1200 dpi laser printer. (We don't even see our graphs with this quality at home—1200 dpi laser printers are expensive!) But do the best you can with the equipment you have—a good picture may be worth more than many words and formulas. If you are asked to "sketch a curve," we mean "draw without a computer." Otherwise we use the word "plot."

2.1   Existence and Uniqueness

Comments

The five Basic Questions posed in this section are at the heart of the theory of differential equations: Does an IVP have a solution? Is it unique? How can a solution be described? How sensitive is a solution to changes in parameters and data? How can a solution be extended; what happens to solutions as time advances? The focus in Chapter 1 has been on obtaining solution formulas. With the formulas in hand, it is often a simple matter to answer the questions. Sometimes solution formulas are too complicated to be of much help in understanding solution behavior. In any case, the questions have to be considered even in the absence of solution formulas. In this section we introduce examples and counterexamples needed to clarify the meaning of the questions, and we show approaches to answering them. When you read this section, you might tie the questions with a specific physical phenomenon that has already been discussed, for example, the "pollution in a tank" problem of Section 1.4. Existence: pollution enters the tank and the pollution level in the tank does build up (so the model IVP for tank pollution levels does have a solution). Uniqueness: if there are two identical tanks, identical amounts of pollutant flowing into the tanks, identical initial amounts of pollutant in the tanks, identical outflow rates, then pollution levels in the tanks are identical (uniqueness of the IVP's solution). Description of solution: words, formulas, pictures, graphs, computer output. Sensitivity/Continuity: change the pollutant levels a little in the inflow stream and the levels in the tank change a little (the solution of the IVP varies continuously with the data). Extension and Long-Term Behavior: after a long time one expects the concentration of pollutant in the tank to be pretty much the same as the concentration of pollutant in the inflow stream (and the formula for the solution of the IVP will mirror that expectation). So the Basic Questions are just the mathematicians' way of encoding in a few words and symbols the everyday experience of people who work with dynamical systems. In addition, the answers to the five Basic Questions are crucial to the use of numerical solvers. Without the Existence, Uniqueness, Extension, and Sensitivity properties, we would never be sure whether the approximation our numerical solver generates has anything to do with the IVP being addressed, nor would we have any idea about its "life-span."

    This section focuses on the existence and the uniqueness of the solution of an IVP. The verifications are given in Appendices A.1 and A.2, although we describe in this section the Picard Iteration scheme which is the key idea of the proof of existence. The Picard idea is simple and we always like to illustrate how it works with a simple example where the iteration generates the Taylor series of the solution.

The question of uniqueness is also central, especially these days when the meaning of chaotic dynamics is a popular topic for scientific debate. If an IVP has an unique solution, then any dynamical system accurately modeled by the IVP is deterministic (not chaotic) in the sense that the present state of the system completely determines its future (and its past) states, at least over the time span where the model is valid. Nonuniqueness introduces an element of uncertainty; which path does a solution follow, and how can the system decide which path to follow? This is a good reason for reading some of the nonuniqueness examples: you might well contrast determinism and uncertainty in this connection. Even when the conditions of uniqueness are satisfied, some systems of ODEs or their numerical approximations appear to have chaotic solutions, but we postpone that discussion to Section 2.7 (numerical approximation) and Chapter 9 (ODEs).

Questions of continuity/sensitivity are left to Section 2.3 and to 2.4, while the use of numerical computer approximations is taken up in Sections 2.5 and 2.6.

1. **(a).** The IVP is $y' = e^t y - y^3$, $y(0) = 0$. The function $f = e^t y - y^3$ is continuous for all t and y since e^t, y, and y^3 are continuous. Moreover, $\partial f/\partial y = e^t - 3y^2$ is also continuous for all t and y. So, the hypotheses of the Existence and Uniqueness Theorem are satisfied in every rectangle, and the given initial value problem has exactly one solution.

   (c). The IVP is $y' = |t||y|$, $y(0) = 1$. The function $f = |t||y|$ and its partial derivative $\partial f/\partial y = |t|$ if $y > 0$, $-|t|$ if $y < 0$ are continuous in any neighborhood of the point $(0,1)$ for which $y > 0$. So, by the Existence and Uniqueness Theorem the initial value problem has exactly one solution in any rectangle containing $(0, 1)$ and lying in the upper half plane $y > 0$.

3. Writing the ODE $ty' - y = t^{n+1}$ in the normal form $y' - y/t = t^n$ and using the integrating factor $1/t$, we have that $(y/t)' = t^{n-1}$. Integrating, $y/t = t^n/n + C$, so we find $y(t) = t^{n+1}/n + Ct$, for any positive n. This formula defines solutions of the original ODE for all t. Since each of these solutions satisfies $y(0) = 0$, the IVP has infinitely many solutions. Since the conditions of the Uniqueness Theorem don't apply at $t = 0$ where $f = y/t + t^n$ is discontinuous, there is no contradiction.

5. **(a).** To solve $y' = y\,\text{step}(t)$, $y(0) = 1$, first suppose $t \geq 0$ and solve $y' = y$, $y(0) = 1$ to obtain $y = e^t$, $t \geq 0$. For $t < 0$, solutions are $y = C$ since $y' = 0$, $t < 0$; set $C = 1$ to obtain the solution $y = 1$ with the property that $\lim_{t\to 0^-} y(t) = \lim_{t\to 0^+} y(t) = 1$. The function

$$y = \begin{cases} e^t, & t \geq 0 \\ 1, & t < 0 \end{cases}$$

is continuous and satisfies the differential equation for $t \geq 0$ (with right-hand derivative $+1$ at $t = 0$) and for $t < 0$. However, this function is not differentiable at $t = 0$ since $\lim_{t\to 0^-} y'(t) = \lim_{t\to 0^-}(y(t) - y(0))/t = \lim_{t\to 0^-}(1 - 1)/t = 0$, while $\lim_{t\to 0^+} y'(t) = \lim_{t\to 0^+}(y(t) - y(0))/t = \lim_{t\to 0^+}(e^t - 1)/t = 1$. By our definition of solution, $y(t)$ is not a solution since it is not differentiable at $t = 0$. (Sometimes it is useful to say that a function such as $y(t)$ is a solution since its right- and left-hand derivatives are equal everywhere except at the point $t = 0$, where they exist but are not equal.) The graph of $y(t)$ has a kink at $t = 0$.

(c). The IVP is $y' = (1 - y^2)^{1/2}$, $y(0) = 1$. The rate function is $f(y) = (1 - y^2)^{1/2}$. Since the partial derivative of f with respect to y, $f_y(0, 1) = -y(1 - y^2)^{-1/2}\big|_{y=1}$ is undefined, the Existence and Uniqueness Theorem does not apply. The constant function $y_1(t) = 1$, all t, is a solution, but the problem may have additional solutions. In fact, if we separate variables, we have $(1 - y^2)^{-1/2}y' = (\arccos y)' = 1$. Integrating, $\arccos y = t + C$, so $y = \cos(t + C)$. Using the initial condition $y(0) = 1$, we can take $C = 0$. Since y' is nonnegative, we restrict t to the interval $-\pi \leq t \leq 0$ [or to any interval of the form $(2n - 1)\pi \leq t \leq 2n\pi$] on which $y = \cos t$ has a nonnegative slope. Notice that the constant function $y(t) = -1$ is also a solution to the ODE (but not the IVP). We can construct another solution of the initial value problem by setting $y = -1$ for $t < -\pi$, then $y = \cos t$ for $-\pi \leq t \leq 0$, and finally $y = +1$ for $t \geq 0$. Such a function is continuous for all t, differentiable for all t (since $y' = -\sin t$ for $-\pi \leq t \leq 0$ and $y'(t) = 0$ for $t \leq -\pi$ and for $t \geq 0$), and satisfies the rate equation and initial condition. In fact, we see that we can build infinitely many solutions by linking the horizontal line $y(t) = -1$ for $t < (2n - 1)\pi$, to the horizontal line $y(t) = 1$ for $t > 2n\pi$, by means of the graph of $y(t) = \cos t$, $(2n - 1)\pi \leq t \leq 2n\pi$, n a nonpositive integer.

7.   **(a).** The equation $ty' + y = 2t$ can be written in normal form as $y' + y/t = 2$ on any t-interval not containing 0. An integrating factor is given by t. The general solution is $y = t + C/t$ for $t \neq 0$. Each value of C gives two distinct solutions, one defined on the t-interval $(-\infty, 0)$ and the other on $(0, \infty)$. Note that $y = t$ is a solution that is defined for all t.

(c). See Fig. 7(c). The solution curve of $y = t$ goes right through the origin. All other solution curves are arcs of hyperbolas and are asymptotic to one end or the other of the y-axis as $t \to 0^+$, or 0^-, and to one end or the other of the line $y = t$ as $t \to \pm\infty$. These properties are implied by Fig. 7(c), and also follow directly from the solution formula.

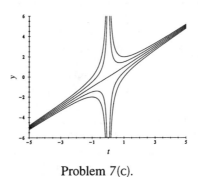

Problem 7(c).

9.   The algorithm for producing the Picard iterates to approximate the solution of the IVP,

$y' = y$, $y(0) = 1$ is

$$y_0(t) = 1, \quad y_1(t) = 1 + \int_0^t y_0(s)\,ds, \quad \dots, \quad y_{n+1}(t) = 1 + \int_0^t y_n(s)\,ds$$

So we have

$$y_1(t) = 1 + \int_0^t 1\,ds = 1 + t$$

$$y_2(t) = 1 + \int_0^t (1+s)\,ds = 1 + t + t^2/2$$

$$y_3(t) = 1 + \int_0^t (1+s+s^2/2)\,ds = 1 + t + t^2/2 + t^3/6$$

$$y_4(t) = 1 + \int_0^t (1+s+s^2/2+s^3/6)\,ds = 1 + t + t^2/2 + t^3/6 + t^4/24$$

Since the Taylor series for e^t is $1 + t + \cdots + t^n/n! + \cdots$ and we know that the solution of the IVP is e^t, we see that the Picard algorithm generates the Taylor series one term at a time.

2.2 Extension and Long-Term Behavior

Comments

In this section we address a topic that is often left out in introductory courses: How far can a solution be extended backward and forward in time? We take up the example IVP, $y' = ky^2$, $y(0) = 1$, where the solution "reaches infinity" in finite time. This can be imagined in the form of a (highly unlikely) physical system modeled by the IVP, where the units of y are, say, miles, time is measured in hours, and the units of k are (miles $\cdot$ hours)$^{-1}$. If $k = 1$, then $y(t)$ is "beyond the end of the universe" within an hour. This also reminds us that a mathematical model has a limited region of usefulness in the space of time and state variables.

Solutions of $y' = f(t, y)$ and the corresponding solution curves cannot just "die" without some reason. A solution can only stop by time T if the solution curve is about to enter a region where the conditions on f given in the Extension Principle fail to hold, or if the solution blows up as time nears the value T (i.e., the solution has a finite escape time). The idea of a maximally extended solution and the existence of periodic forced oscillations are two important topics in this section. We like Example 2.2.7 on the transmission of a "message." We like to emphasize the long-term behavior of solutions of an autonomous ODE $y' = f(y)$: Every solution either tends to $\pm\infty$ or to an equilibrium solution as t increases to $+\infty$. This question of the long-term behavior of solutions of autonomous ODEs is a hot issue today, because it is directly related to chaos. We come back to the question in Chapter 9 for a system of autonomous ODEs. We explain in Section 9.2 just what the long-term behavior of a bounded solution of a planar autonomous system can be (chaos cannot occur in dimension 2, just as Theorem 2.2.2 says it can't occur in dimension 1). However, chaos can occur if an autonomous system has 3 state variables (see the Lorenz system of Section 9.4). What we just said applies to ODEs, but *not* to discrete dynamical systems. As we note in Section 2.7, chaotic wandering *is* possible with a single state variable if we are dealing with certain discrete one-dimensional transformations.

1. **(a).** Writing the ODE $yy' = -t$ as $(y^2)'/2 = -t$ and then integrating, we have $y^2/2 = -t^2/2 + C$, so $y = \pm\sqrt{2C - t^2}$. Using the initial condition $y(0) = 1$, we find $C = 1/2$ and we take the plus sign. So the maximally extended solution satisfying $y(0) = 1$ is $y = \sqrt{1 - t^2}$ and the maximal interval of existence of this solution is $-1 < t < 1$. At $t = \pm 1$, the solution's slope would be infinite. See Fig. 1(a) for the graph of the maximally extended solution. Note that if we were to try to extend the time interval to values of t for which $|t| > 1$, then y and y' would both be complex-valued.

Problem 1(a).

3. Separating variables in the ODE, $y' = 1 - y^2$, and integrating, we obtain the general solution

$$\ln|(1 + y)/(1 - y)|^{1/2} = t + C, \quad \text{where } C \text{ is an arbitrary constant}$$

Exponentiating, squaring, dropping the absolute value signs, and redefining the constant of integration we have

$$\frac{1 + y}{1 - y} = ce^{2t}, \quad \text{where } c \text{ is an arbitrary constant} \tag{i}$$

Imposing the initial condition $y(0) = y_0$, we see that $c = (1 + y_0)/(1 - y_0)$. Solving (i) for y we obtain

$$y = \frac{c - e^{-2t}}{e^{-2t} + c}$$

Now if $y_0 < -1$, then we have that $-1 < c < 0$, and $y(t)$ escapes to $-\infty$ as $t \to t_e^-$, where $e^{-2t_e} + c = 0$. Solving $e^{-2t_e} + c = 0$ for the escape time t_e,

$$t_e = (-1/2)\ln(-c) = (-1/2)\ln\frac{1 + y_0}{y_0 - 1}$$

5. Recall that $y = y_0$ is an equilibrium solution of $y' = f(y)$ if $f(y_0) = 0$.

(a). The equilibrium solutions of $y' = (1 - y)(y + 1)^2$ are $y = \pm 1$. See Fig. 5(a). The solutions between $y = -1$ and $y = +1$ tend to $y = 1$ as $t \to +\infty$ and to $y = -1$ as $t \to -\infty$.

(c). The equilibrium solutions of $y' = y(y - 1)(y - 2)$ are $y = 0$, $y = 1$, and $y = 2$. See Fig. 5(c). Solutions between $y = 1$ and $y = 2$ tend to $y = 1$ as $t \to +\infty$ and to $y = 2$ as $t \to -\infty$. Solutions between $y = 0$ and $y = 1$ tend to $y = 1$ as $t \to +\infty$ and to $y = 0$ as $t \to -\infty$.

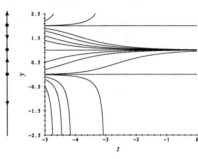

Problem 5(a).   Problem 5(c).

7.   Solution curves rise on one side of a nullcline (and fall on the other) in all of these graphs. The nullclines in the graphs are dashed.

(a). The ODE is $y' = (y + 3)(y - 2)$. See Fig 7(a). The nullclines are $y = -3$, $y = 2$. Notice that the nullclines are also solutions.

(c). The ODE is $y' = ty - 1$. See Fig 7(c). The nullcline is $y = 1/t$.

(e). The ODE is $y' + (\sin t)y = t \cos t$. See Fig 7(e). The nullcline is $y \sin t = t \cos t$.

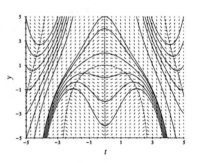

Problem 7(a).   Problem 7(c).   Problem 7(e).

9.   **(a).** The ODE $y' = y^2$ separates to $y^{-2} dy = dt$. Integrating, $-1/y = t + C$ so if $y(0) = 5$, then $C = -1/5$ so $-1/y(t) + 1/5 = t$, and $y = 5/(1 - 5t)$, where we require that t be less than $1/5$. As $t \to (1/5)^-$, $y(t) \to +\infty$. As $t \to -\infty$, $y(t) \to 0^+$.

11. All of the solution curves [except for part **(c)**] tend to a periodic forced oscillation (Theorem 2.2.4) because the coefficient p_0 in every linear ODE, $y' + p_0 y = q(t)$, is positive [except for part **(c)**]. This means that the term $e^{-p_0 t} C$ in the general solution of the ODE tends to 0 as $t \to +\infty$, leaving the periodic forced oscillation as the only visible solution. The solutions of the ODE in part **(c)** are $y = C + \int_0^t 3 \, \mathrm{sqw}(s, 50, 2) \, ds$; since the integrand has value 3 over half of every time interval of length 2 (and is 0 otherwise), the value of the integral increases piecemeal. The constant C simply translates the solution curves up or down according to the sign of C. The corners on all of the solution curves except parts **(e)** and **(f)** reflect the discontinuities in the driving terms. The graphs for parts **(e)** and **(f)** are smooth since the driving terms are continuous. The ODEs and some of their solution curves are shown below.

(a). $y' + y = 3 \, \mathrm{sqw}(t, 25, 3)$,

(c). $y' = 3 \, \mathrm{sqw}(t, 50, 2)$,

(e). $y' + y = \sin 3t$,

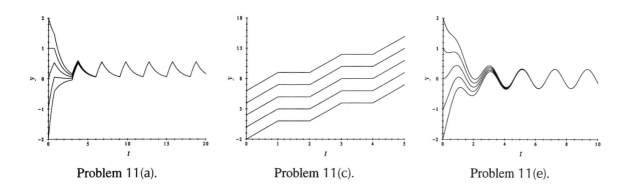

Problem 11(a).   Problem 11(c).   Problem 11(e).

13. Group project.

Background Material: Sign Analysis and Long-Term Behavior

Here is a verification of part of Theorem 2.2.3. Suppose that $y_1 < y_2$ are two consecutive zeros of $f(y)$; that is, $f(y_1) = f(y_2) = 0$, but $f(y)$ does not vanish in the interval $y_1 < y < y_2$. Then since $f(y)$ is continuous, $f(y)$ must be of one sign in the strip $y_1 < y < y_2$, say positive. Note that the lines $y = y_1$ and $y = y_2$ are solution curves for the ODE. Now, because f is positive, the solution that passes through the point (t_0, y_0), $y_1 < y_0 < y_2$, must be increasing as t increases, if the solution curve remains in the strip $y_1 < y < y_2$. To leave this strip as t increases, the solution curve would have to cross the equilibrium solution curve $y = y_2$, but the Uniqueness Theorem implies that this cannot happen. The Extension Theorem easily shows that our solution curve is defined for all $t \geq t_0$. Similarly, we see that for backward time our solution curve falls for decreasing t, but it never crosses the equilibrium solution $y = y_1$. The Extension Theorem shows again that the solution is defined for all $t \leq t_0$. If follows that the limits

$$\lim_{t \to +\infty} y(t) = b \leq y_2 \quad \text{and} \quad \lim_{t \to -\infty} y(t) = a \geq y_1$$

both exist. We now show that $a = y_1$ and $b = y_2$. By shifting this solution curve in t we can completely fill up the strip $a < y < b$ with the solution curves of the ODE. Now suppose that $b < y_2$, contrary to our assertion. Then, as noted above, the solution curve defined by the IVP $y' = f(y)$, $y(t_0) = b$, can be extended to the entire t-axis. Extending the solution forward, the solution must cross the line $y = b$ because $f(b) > 0$, and so must intersect a solution curve in the strip $a < y < b$, in contradiction to the Uniqueness Theorem. It must be that $b = y_2$. Similarly, it can be shown that $a = y_1$.

2.3 Sensitivity

Comments

Information about how the solution of an IVP changes as the data change is critical for a real understanding of the IVP, particularly if the IVP models some real phenomenon. The idea of sensitivity is fairly easy to grasp, even though we never give a precise definition. The informal explanation of the solution of an IVP being a continuous function of all the data is also easy to understand, but it must be emphasized that the time interval about t_0 must be small. Note that small changes in data may evolve over time into huge variations. This is good background for a later discussion on the Lorenz system (Section 9.4), where we speak of chaos. We only treat linear ODEs in this section. See Appendix A.4 for a discussion of the sensitivity question for the IVP, $y' = f(t, y)$, $y(t_0) = y_0$.

1. The solution curves for the IVPs $y' = -y \sin t + ct \cos t$, $y(-4) = -6$, $c = 0.7, 0.8, \ldots,$ 1.3 in Fig. 2.3.1 in the text seem to be diverging as $t \to -1^-$, but the plot in Fig. 1 below for $-4 < t < 20$ shows that they alternately spread out and bunch together. It appears that as t increases, the curves oscillate with ever-increasing amplitude, after a while bunching up at the peaks and spreading out at the troughs, with successively larger spreads at each trough. This pattern is only our conjecture. Do you have a different conjecture?

Problem 1.

3. We are given that the vat has a 100 gallon capacity, that r gallons of brine run into the vat per minute, and the salt concentration in the inflow stream is $c(t)$ lbs/gal. The brine exits the tank at the rate of r gal/min. Let $S(t)$ be the lbs of salt dissolved in the brine in the tank at time t. We are given that $S(0) = 5$. Then $S'(t) =$ Rate in $-$ Rate out $= rc(t) - rS/100$, $S(0) = 5$, where, say, $0 < r_0 \leq r \leq r_1$ and $0 \leq c(t) \leq c_0$. The Bounded Input–Bounded Output Principle applies to the ODE, $S' + rS/100 = rc$. Identifying $|y_0| = 5$, $p_0 = r_0/100$, and $M = r_1 c_0$, we have that $|S(t)| \leq 5e^{-r_0 t/100} + (100/r_0)r_1 c_0(1 - e^{-r_0 t/100}) \leq 5 + 100 r_1 c_0/r_0$. Since $c(t) = S(t)/100$, the safety rule that $c(t)$ must never exceed 0.1 lb/gal implies that we must be sure that $S(t) \leq 10$ at all times, which we can guarantee by setting $100 r_1 c_0/r_0 \leq 5$, that is, we require that $20 r_1 c_0 \leq r_0$ to ensure that the concentration of salt in the tank never exceeds 0.1 lbs/gal.

5. **(a).** Using the integrating factor e^t to solve $y' = -y + e^{-t}$, we see that $(ye^t)' = 1$, so integrating and using $y(0) = a$, we have $y(t, a) = ae^{-t} + te^{-t}$; similarly for $y(0) = b$. We have

$$|y(t, a) - y(t, b)| = |a - b|e^{-t}, \quad t \geq 0$$

The solution is relatively insensitive to changes in initial data because as time goes on, the deviation between the solutions decays to zero.

(c). The solution $y_c(t, a)$ of the IVP $y' = -y + ct$, $y(0) = a$ is obtained by using the integrating factor e^t and then integration by parts (or a table of integrals): $y_c(t, a) = ae^{-t} + c(e^{-t} + t - 1)$. Using this formula and the Triangle Inequality ($|A + B| \leq |A| + |B|$), we see that

$$|y_c(t, a) - y_1(t, b)| \leq |a - b|e^{-t} + |c - 1| \, |e^{-t} + t - 1|$$

As t increases from 0 to 10, the first term on the right decays to a small fraction of $|a - b|$, but the last term has increased quite a bit. So, on the interval $0 \leq t \leq 10$ the solution is insensitive to the changes in the initial data, but highly sensitive to the changes in the parameter c.

7. The solution $y(t)$ of the IVP, $y' + p_0 y = M$, $y(0) = y_0$, is $y = y_0 e^{-p_0 t} + M(1 - e^{-p_0 t})/p_0$; that is, if y_0 and p_0 are positive, then y precisely equals the right-hand side of the inequality given in the Bounded Input–Bounded Output Principle. We have shown that for the specific case $p(t) = p_0$, $q(t) = M$, the inequality can't be improved. So, it is not in general possible to replace the right-hand side by a smaller quantity.

9. The ODE $y' + y = t$ has integrating factor e^t, and its solutions are given by $y = Ce^{-t} + t - 1$. So $y \to +\infty$ as $t \to +\infty$. Note that the input t is not bounded on the interval $t \geq 0$. We have shown that for this specific ODE, $q(t)$ is unbounded and BIBO fails. The condition $|q(t)| \leq M$ cannot in general be dropped.

Background Material: Verification of Theorem 2.3.1

Here is the verification of Theorem 2.3.1 that was omitted from the text. Suppose that $p(t) \geq p_0$ where $p_0 \neq 0$. Then by properties of integrals and exponentials, we have for all positive values of t that

$$p_0 t \leq \int_0^t p(s)\, ds$$

$$e^{-\int_0^t p(s)\, ds} \leq e^{-p_0 t}$$

Since the solution of the IVP

$$y' + p(t)y = q(t), \quad y(0) = y_0$$

is given by

$$y(t) = e^{-\int_0^t p(s)\, ds} \left\{ y_0 + \int_0^t e^{\int_0^s p(r)\, dr} q(s)\, ds \right\}$$

$$= e^{-\int_0^t p(s)\, ds} y_0 + \int_0^t e^{-\int_s^t p(r)\, dr} q(s)\, ds$$

we have [using the triangle inequality and the fact that $|\int_s^t p(r)\, dr| \leq \int_s^t |p(r)|\, dr$ if $t \geq s$]:

$$|y(t)| \leq e^{-p_0 t} |y_0| + \int_0^t e^{-p_0(t-s)} M\, ds$$

$$\leq e^{-p_0 t} |y_0| + \frac{M}{p_0} e^{-p_0 t} \left[e^{p_0 t} - 1 \right]$$

$$\leq e^{-p_0 t} |y_0| + \frac{M}{p_0} \left[1 - e^{-p_0 t} \right]$$

$$\leq |y_0| + \frac{M}{p_0} \quad \text{(for } t \geq 0)$$

This verifies Theorem 2.3.1.

2.4   Introduction to Bifurcations

Comments

Bifurcations are new to the introductory ODE syllabus, but they are an important part of any course that introduces the ideas of sensitivity to data changes and is serious about modeling physical systems. Although most changes in the data do not lead to drastic changes in solution behavior (at least not over a short span of time), there are those occasional data ranges where something completely new occurs, although the consequences may only become evident with time's advance. The term "bifurcation" is now applied to this kind of event, but it is not a precisely defined term. We introduce this idea via examples and suggestive (rather than exact) definitions. We introduce the saddle-node bifurcation in the context of a harvested/restocked and logistically changing population. We do the important task of rescaling so that the only remaining parameter in the model ODE is the harvesting/restocking rate. This is a good example to see just why rescaling is often done before computing. In reading this material, you might think of four stages after the scaling has been done. First, find the equation of the equilibrium solutions in terms of the parameter c and plot and discuss the meaning of their graphs in the cy-plane (the bifurcation diagram), clearly locating the bifurcation point. Next, show graphs of solutions before, at, and beyond bifurcation. Then replot the bifurcation diagram, but this time introducing solid or dashed arcs depending upon whether an equilibrium solution is an attractor or a repellor. Finally, plot the rate function $f = f(y, c)$ in a yf-plane for various values of c below, at, and above the bifurcation value, and show that at the bifurcation value c^* the graph of $f = f(y, c^*)$ is tangent to $f = 0$. You may want to complete these stages in a different order.

1. Find the equilibrium solutions y_i as functions of the parameter c. Then find the value of c where the complex equilibrium solutions merge into a single real equilibrium solution and then split into two real equilibrium solutions. The sign of the rate function near an equilibrium determines whether the equilibrium is an attractor or a repeller. For example, if $f(y_1, c_1) = 0$ and if $f(y, c_1) > 0$ for y near y_1 but $y < y_1$, while $f(y, c_1) < 0$ for y near y_1 and $y > y_1$, then $y = y_1$ is an attracting equilibrium for $c = c_1$ because nearby solutions rise toward it or fall toward it as time increases. When drawing the bifurcation diagram in the cy-plane, use a solid arc for the attracting branch and a dashed arc for the repelling branch.

(a). The ODE is $y' = c - y^2$. Since $f(y, c) = c - y^2$, if $c > 0$ the equilibrium solutions are at $y = \pm\sqrt{c}$; if $c = 0$, $y = 0$ is the only equilibrium solution, while if $c < 0$, there are no real equilibrium solutions since $\sqrt{c}$ is complex. So $c = 0$ is the critical value. As c increases through 0, the complex conjugate numbers $\pm\sqrt{c}$ move toward each other in the complex plane, meet when $c = 0$, and split apart into a pair of real equilibrium solutions. The solution $y = \sqrt{c}$, $c > 0$, is an attractor since if $y < \sqrt{c}$ but near $\sqrt{c}$, then $f = c - y^2$ is positive and solutions of $y' = c - y^2$ rise up toward $y = \sqrt{c}$, while if $y > \sqrt{c}$, then $c - y^2$ is negative and solutions fall toward the equilibrium. A similar argument shows that $y = -\sqrt{c}$ is a repeller. See Fig. 1(a) for the bifurcation diagram.

(c). The ODE is $y' = c + 2y + y^2$. The zeros of $f(y, c) = c + 2y + y^2$ are $y_{1,2} = -1 \mp \sqrt{1 - c}$, and are real and distinct for $c < 1$, real and identical for $c = 1$, and complex conjugates for $c > 1$. For large y, $f(y, c)$ is positive, but for $y < y_2$ and near y_2, $f(y, c)$ is negative. So $y = y_2$ is a repeller. A similar argument shows that $y = y_1$ is an attractor. See Fig. 1(c) for the bifurcation diagram.

Problem 1(a).

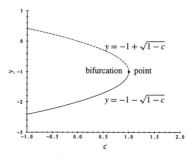

Problem 1(c).

3. First find the equilibrium solutions by finding the zeros of each rate function as functions of the parameter c. Then find the value of c where the zeros of the rate function merge into a single real zero of the rate function. For c on one side of this value the rate function in each part of this problem has a single real zero. For c on the other side of the value the rate function has three real zeros. The bifurcation diagram shows the real zeros as functions of the parameter c. The solid arcs denote attracting equilibrium solutions; the dashed arcs repelling solutions. For each value c_i of c each real equilibrium solution y_i is shown to be an attractor [repeller] if for values of y near y_i, $y > y_i$, the sign of $f(y, c_i)$ is negative [positive] and for $y < y_i$ the sign is positive [negative]. In the pitchfork bifurcation diagrams, arcs of attracting equilibrium solutions are solid, while arcs of repelling equilibrium solutions are dashed. See the text subsection on the pitchfork bifurcation for more details.

    (a). The ODE is $y' = (c - 2y^2)y$. The zeros of $f = (c - 2y^2)y$ are $y_1 = 0$, $y_{2,3} = \mp\sqrt{c/2}$, and the latter two are real for $c > 0$, merge with y_1 for $c = 0$, and are complex for $c < 0$. For $c > 0$, y_2 and y_3 are attractors since $f(y, c)$ is negative for y near y_2 [y_3] if $y > y_2$ [$y > y_3$] and positive if $y < y_2$ [$y < y_3$]. For $c > 0$, y_1 is a repeller since $f(y, c)$ is positive for y near 0, $y > 0$, and negative for $y < 0$. For $c < 0$, y_1 is an attractor since $f(y, c)$ is negative for $y > 0$ and positive for $y < 0$. See Fig. 3(a).

    (c). The ODE is $y' = (c - y^4)y$. The five zeros of $f = (c - y^4)y$ are $y_1 = 0$, $y_{2,3} = \mp c^{1/4}$, and $y_{4,5} = \mp i|c|^{1/4}$. The only zeros of interest here are y_1, y_2, y_3 because $y_{4,5}$ are complex, except if $c = 0$ in which case all five coincide at $y = 0$. The zeros $y_{2,3}$ are real for $c > 0$ where they are both attractors because $f(y, c)$ is negative for y near y_2 or y_3 with $y > y_2$ or $y > y_3$, but positive for $y < y_2$ or $y < y_3$. The zero y_1 is a repeller [attractor] for $c > 0$ [$c < 0$] because $f(y, c)$ is positive [negative] for y near 0 but $y > 0$ and negative [positive] for $y < 0$. See Fig. 3(c).

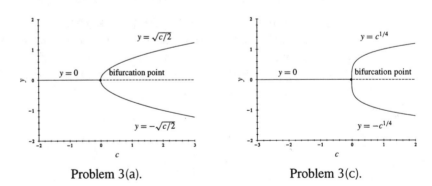

Problem 3(a).   Problem 3(c).

5. In each case set the rate function $f(y, c)$ equal to zero to find the equilibrium solutions. One of these is $y_1 = 0$, the other, y_2, is a function of c and real for all c. As c crosses a critical value, y_2 moves toward y_1, merges with it at the critical value, and then moves away. Except at the critical value of c where the transcritical bifurcation occurs, one of y_1 and y_2 is an attractor and the other a repeller [because of the signs of $f(y, c)$] but as c

crosses the bifurcation value the attractor becomes a repeller and vice-versa. The transcritical bifurcation diagram in the cy-plane shows the graphs of $y_1 = 0$ and y_2 as a function of c. Solid arcs denote an attractor, dashed arcs a repeller. As noted above, the signs of $f(y, c)$ near an equilibrium value of y are used to determine whether the equilibrium is an attractor or a repeller. At the bifurcation value of c, the equilibrium solution curve attracts on one side and repels on the other.

(a). The ODE is $y' = cy - y^2$. Here $f(y, c) = cy - y^2 = y(c - y)$, so $y_1 = 0$ and $y = c$ are the two equilibrium curves and the transcritical bifurcation occurs for $c = 0$. See Fig. 5(a), Graphs 1–3 for solution curves for $c = -1$ ($y_1 = 0$ is an attractor and $y_2 = -1$ is a repeller), $c = 0$ at the transcritical bifurcation value ($y = 0$ attracts on one side and repels on the other), $c = 1$ ($y_1 = 0$ is a repeller and $y_2 = 1$ is an attractor). Figure 5(a), Graph 4 shows the bifurcation diagram in the cy-plane: the two lines are $y = 0$, $y = c$. Observe how the attractor switches to a repeller, and vice versa as c crosses the bifurcation value 0.

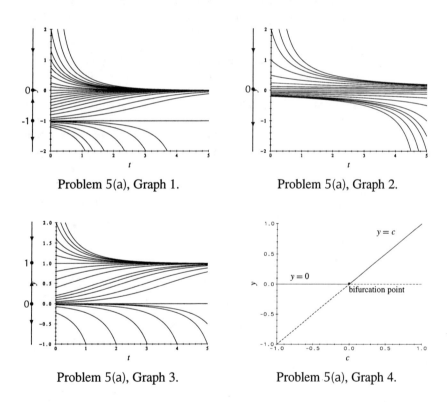

Problem 5(a), Graph 1.   Problem 5(a), Graph 2.

Problem 5(a), Graph 3.   Problem 5(a), Graph 4.

7. The ODE is $P' = (1 - P/1000)P - H$.

(a). Suppose that N hunting licenses are issued and (in the worst case scenario from the duck's viewpoint) each hunter bags the maximum of 20 ducks. Measuring time in years, the model ODE is $P' = (1 - P/1000)P - 20N$. The zeros of the rate function are given by using the quadratic formula

$$P_{1,2} = 500 \mp (10^6 - 8 \cdot 10^4 N)^{1/2}/2$$

So we must set N below the critical value which is where the quantity inside the square root moves from positive to negative (i.e., where it is zero). That value is $10^6 - 8 \cdot 10^4 N = 0$, or $N = 10^2/8 = 12.5$. So no more than 12 licenses should be issued.

2.5 Approximate Solutions

Comments

How does a numerical solver approximate the solution of the first-order IVP $y' = f(t, y)$, $y(t_0) = y_0$? The underlying principle of all solvers is simple enough: approximate the rate function at a point, multiply by a time increment, and repeat. The differences in the various methods have to do with how they approximate the rate function and how the step size is adjusted. In addition, many solvers have adjustable controls on allowable absolute and relative errors, minimum and maximum internal step size (for adaptive methods), number of points plotted, and so on. This section is a brief introduction to a complex and fascinating area of numerical analysis—the search for the "best" ODE solver. Of course, no such thing exists since for every numerical solver yet invented there is some IVP that "breaks" it. Nevertheless, there are some basic notions that can be conveyed about solvers. We focus on one-step algorithms such as the Euler Method and Runge-Kutta Method, and use examples to illustrate what they can do. Actual numerical estimates are given below only for the Euler and RK4 methods. An excellent source for a more extensive treatment is L.F. Shampine's book, *Numerical Solutions of Ordinary Differential Equations*, Chapman & Hall, New York, 1994.

1. Approximate values for $y(1)$, where $y' = -y$, $y(0) = 1$, with step $h = 0.1$, will vary according to the calculator or numerical solver used. The exact answer is $1/e = 0.367879441...$. See Fig. 1 for Euler's Solution (dashed), RK4 (solid).

    (a). Euler's algorithm, $y_n = y_{n-1} - hy_{n-1}$, $1 \leq n \leq 10$, gives $y(1) \approx y_{10} = 0.348678$ (first six decimals).

    (c). Fourth-order Runge-Kutta, $y_n = y_{n-1} + h(k_1 + 2k_2 + 3k_3 + k_4)/6$, where $1 \leq n \leq 10$, $k_1 = -y_{n-1}$, $k_2 = -(y_{n-1} + hk_1/2)$, $k_3 = -(y_{n-1} + hk_2/2)$, $k_4 = -(y_{n-1} + hk_3)$, gives $y(1) \approx y_{10} = 0.367879$ (first six decimals).

Problem 1.

3. Euler's method [part **(a)**] and RK4 [part **(b)**] are used to approximate $y(1)$ if $y' = -y^3 + t^2$, $y(0) = 0$, and the step size h is 0.1, 0.01, 0.001.

(a). The first six decimals of the respective Euler approximations to $y(1)$ for the three values of h are 0.283786, 0.325058, and 0.329236. Figure 3(a) shows the three Euler polygons.

Problem 3(a).

5. We shall interpret a as t_j and b as t_{j+1} in an RK4 process for solving $y' = f(t)$, $y(t_0) = y_0$ over an interval containing a, b, $a < b$. Suppose y_j has been calculated as an estimate for $y(t_0 + jh)$, where $y'(t) = f(t)$, $y(t_0) = y_0$, $t_0 \leq t \leq T$. Then $y_{j+1} = y_j + \int_{t_j}^{t_{j+1}} f(t)dt \approx y_j + (h/6)[f(t_j) + 4f(t_j + h/2) + f(t_{j+1})]$ by Simpson's formula, where $h = t_{j+1} - t_j$. Fourth-order Runge-Kutta gives $y_{j+1} = y_j + (h/6)[k_1 + 2k_2 + 2k_3 + k_4]$, where $k_1 = f(t_j)$, $k_2 = f(t_j + h/2) = k_3$, $k_4 = f(t_j + h) = f(t_{j+1})$. So, Simpson's method and fourth-order Runge-Kutta yield identical values for y_{j+1}, once y_j has been calculated.

2.6 Computer Implementation

Comments

Given any numerical algorithm for solving ODEs, there will be some IVP for which the algorithm works poorly, if at all. This is a fact of numerical algorithms, so we address here a few of the things that can go wrong when using a solver. Problems with numerical algorithms show how important it is not to place unquestioning faith in their reliability. By this point in the text we hope that the reader will have begun to see that each of the three approaches to ODEs (finding solution formulas, using theory, computing numerical approximations) has both advantages and drawbacks. Understanding the pluses and minuses of each approach contributes to a deeper understanding of the behavior of the ODEs themselves.

Tips on using numerical solvers

In Section 2.2 we talked about extending a solution by taking the final values of the state variables and using them as initial data for the extended solution. There are at least two reasons why this procedure is not recommended when using numerical solvers:

- The solver's user interface may report numerical data with less precision than the solver itself uses for calculations (i.e., the solver may report and accept data with 4 digits of precision but use 15 digits of precision internally. Look at the ODE $y' = 3y \sin y + t$ some of whose solutions appear in the chapter cover figure. Imagine that your solver produces an approximate value for $y(t)$ at $t_{final} = -6$ and that you wish to extend your solution forward in time. Can you see the trouble you would have if y_{final} is very near 0?

- Some solvers are adaptive in that they make decisions that affect the solver engine at every time step. These decisions are based on the recent history of the solution process. This is especially true of solvers based on multistep methods like the one used to produce the figures in this text. If the solution process is restarted at one point, then the history is lost and the solver is left to its own devices on how to start the solution process. Should this happen at a sensitive solution point like the one mentioned above, then some surprises may be in store for the user.

1. **(a).** Separating variables in the ODE, $y' = y^3$, we have $(y^{-3})' = dt$. Integrating we get $-y^{-2}/2 = t + C$ Using the initial condition $y(0) = 1$ we get, $y^{-2} = 1 - 2t$, that is, $y = (1 - 2t)^{-1/2}$. Therefore, the t-interval on which this maximally extended solution is defined is $(-\infty, 1/2)$.

    (c). See Fig. 1 for the graphs of $y(t)$ (dashed) (actually Runge-Kutta with step 0.0001) and the Euler Solution (solid curve). The Euler Solution stepped across the singularity at $t = 1/2$ and started tracking a neighboring solution curve.

Problem 1.

3. **(a).** To solve $y' + 2y = \cos t$, we use the integrating factor e^{2t} and a table of integrals (or integration by parts). We see that $(ye^{2t})' = e^{2t} \cos t$. Integrating, $ye^{2t} = \int e^{2t} \cos t \, dt + C$, so $y = Ce^{-2t} + 0.4 \cos t + 0.2 \sin t$, where C is an arbitrary constant. As $t \to +\infty$, $y(t) \to y_p(t) = 0.4 \cos t + 0.2 \sin t$.

    (c). Fig. 3(c) is the forward solution curve of $y' + 2y = \cos t$, $y(0) = 0.4$ as approximated by RK4 with step size $h = 0.1$; the computed approximation to $y(20)$ is $0.34582020\ldots$.

    (e). See Fig. 3(e), Graphs 1 and 2, corresponding to $h = 0.01, 0.001$. The problem with going backward in time is that the forward solutions converge very rapidly to the particular solution $y_p = 0.4 \cos t + 0.2 \sin t$, but numerical solvers can only give approximate answers.

Thus, $y(20) \approx y_p(20)$ can only be computed approximately, no matter how small a step size is used. In effect, when the backward solutions are attempted, they start at the different computed approximations corresponding to 0.01 and 0.001. While solving backwards, small errors occur. Due to sensitivity, these cause the failure to accurately solve backwards. The computed values of the approximations at $t = 20$ are printed on the graphs.

| Problem 3(c). | Problem 3(e), Graph 1. | Problem 3(e), Graph 2. |

5. **(a).** The IVP is $y' = (1 - y)y$, $y(0) = y_0$. There are two equilibrium solutions, $y(t) = 1$ and $y(t) = 0$, corresponding to initial conditions $y_0 = 1$ and $y_0 = 0$, respectively. Separating variables in the ODE, $y' = y(1 - y)$, we have $[1/y + 1/(1 - y)]y' = 1$, so $\ln|y/(1 - y)| = t + C$. Since $y(0) = y_0$, $C = \ln|y_0/(1 - y_0)|$. So

$$\ln|y/(1 - y)| = \ln|y_0/(1 - y_0)| + t$$

$$\ln\left|\frac{y(1 - y_0)}{(1 - y)y_0}\right| = t$$

$$\frac{y(1 - y_0)}{(1 - y)y_0} = e^t$$

Solving for y,

$$y = \frac{y_0 e^t}{1 - y_0 + y_0 e^t} = \frac{y_0}{y_0 + (1 - y_0)e^{-t}}, \qquad t_0 > 0$$

As $t \to +\infty$, $y \to 1$.

(c). If $h = 1.5$, then $y_n = y_{n-1} + 1.5y_{n-1}(1 - y_{n-1})$, $n = 1, 2, \ldots, 6$, with $y_0 = 1.4$. See Fig. 5(c), where the Euler Solution is the solid curve. The first segment of the Euler Solution again goes way below the equilibrium solution $y = 1$, but, because the step size $h = 1.5$ is so large, some of the later segments also overshoot. It appears that the Euler Solution oscillates about the equilibrium, $y = 1$, but approaches equilibrium as n increases.

Problem 5(c).

7. **(a).** The rate function is $-y(1-y)^2$. Since $(1-y)^2$ is positive for $y \neq 1$, y' is negative for $y > 0$, so for $y_0 > 1$, solution curves of the ODE approach the equilibrium solution $y = 1$ from above, whereas for $y_0 < 1$ solutions diverge from the equilibrium $y = 1$ and fall toward the equilibrium $y = 0$. So, $y = 1$ is semistable, attracting solutions from above and repelling solutions from below.

Background Material: Effect of Round-off Error

Suppose Euler's Method is used to estimate $y(3)$, where $y(t)$ solves the IVP,

$$y' = y, \qquad y(0) = 0$$

The plot below shows the error in approximating $y(3)$ as a function of the total number of steps.

Total number of steps.

2.7 Euler's Method, the Logistic ODE, and Chaos

Comments

Section 2.7 takes up the application of a single numerical algorithm (Euler's Method with step size

h) and what happens when it is applied to approximate solutions of the logistic ODE, $y' = (1 - y)y$. The results are startling and show how sensitive the Euler algorithm, $y_n = y_{n-1} + h(1 - y_{n-1})y_{n-1}$, is to changes in the step size. You may want to ignore ODEs entirely and approach the Euler Method as a discrete dynamical system that depends on the parameter h. From this viewpoint, we see period doubling sequences of values of h, sequences that are now known to lead to chaotic wandering. However, a single section on the topic of chaotic wandering and discrete dynamical systems can hardly do justice to this fascinating topic. For this reason we have listed in the text (and also in the Background Material below) references for further reading in the area. **Warning:** You can easily get hooked on the graphical images of chaotic wandering, to the detriment of all other topics in the syllabus. We shall return to chaos in Section 9.4 when we consider the Lorenz system. We have avoided giving a mathematical definition of chaos and have opted instead to use the descriptive term, chaotic wandering. There is a good reason for avoiding the technical definition—there is as yet no agreement on just what the definition should be! See Devaney [*An Introduction to Chaotic Dynamical Systems*, Addison-Wesley, New York, 1989] for one definition and M. Vellekoop and R. Berglund [*Amer. Math. Monthly*, April 1994, pp. 353–355] for comments on Devaney's definition. We describe the bifurcation diagrams (Figs. 2.7.7, 2.7.8) for the Euler algorithm as h increases.

The book by James Gleick, *Chaos*, Viking Penguin Inc., Harrisonburg, VA, 1987, tells the story of how the modern developments in chaos came about.

---

1. Let u be a value of y_0 giving rise to a 2-cycle in the recursion relation $y_{n+1} = y_n + hy_n(1 - y_n)$ and put $v = u + hu(1 - u)$. Then for u to give rise to a 2-cycle, we must have $u = v + hv(1 - v)$. That is, $y_k = u$ implies $y_{k+1} = v$ implies $y_{k+2} = u$ since u gives rise to a 2-cycle. We can plug the expression for u in terms of v into the expression for v in terms of u. So, $v = v + hv(1 - v) + h[v + hv(1 - v)][1 - v - hv(1 - v)]$, or (after some algebra), $hv(1 - v)[h^2v^2 - (2h + h^2)v + h + 2] = 0$. If $v = 0$ or $v = 1$, then $y_n = 0$ for all n, or $y_n = 1$ for all n in which case we have a constant solution rather than a 2-cycle. Otherwise, v must be a root of the quadratic, $h^2v^2 - (2h + h^2)v + h + 2 = 0$, which has only the two roots $v_{1,2} = (2 + h \pm \sqrt{h^2 - 4})/(2h)$. There are only these 2 values of y_0 in $0 < y_0 < \frac{(1+h)}{h}$ that give rise to 2-cycles.

3. **(a).** Figure 3(a), Graphs 1–4 show the segments of the Euler Solution for the IVP, $y' = y(1 - y)$, $y(0) = 1.2$, $0 \le t \le 400$ with $r = 1$, $h = 2.65$. The segments are plotted over the subintervals $[0, 100]$, $[100, 200]$, $[200, 300]$, $[300, 400]$. There appear to be no cycles and no particular pattern—and so chaotic wandering about the equilibrium $y = 1$.

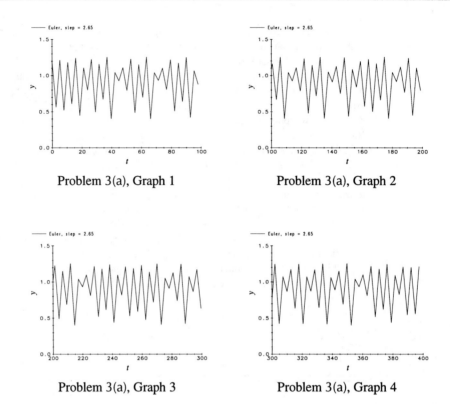

Problem 3(a), Graph 1   Problem 3(a), Graph 2

Problem 3(a), Graph 3   Problem 3(a), Graph 4

5. Group project. The plot below illustrates the geometry of the parabola given by plotting y_n versus y_{n-1}.

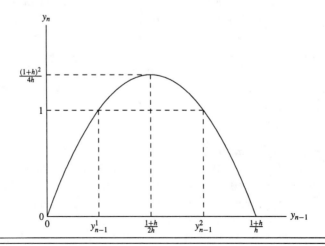

Background Material: A 32-cycle of the logistic equation

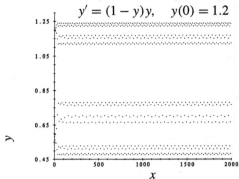

$$y' = (1 - y)y, \quad y(0) = 1.2$$

Approach to the Euler points of a 32-cycle Euler Solution: $h = 2.569$.

The figure above shows that the Euler Solution with $h = 2.569$ and $y_0 = 1.2$ tends to a 32-cycle Euler Solution, but the vertical separation of some of the lines of Euler points is so small that distinct levels are hard to see. In fact, each of the sixteen visible levels actually has two lines, but only zooming will split these into visibly different lines.

Second-Order
Differential Equations

The up-and-down motion of a body suspended by a spring may be modeled by a second-order ODE (Section 3.1). Why second-order? Like most dynamical systems where forces are involved, Newton's Second Law comes into play:

$$\text{mass} \times \text{acceleration of body} = \text{sum of forces acting on body}$$

Acceleration is the second derivative with respect to time of the state variable representing the body's position. The proper use of Newton's Laws involves vectors, but the motion here is only one-dimensional (along the local vertical under the point of suspension of the spring), and we can get away with real-valued state variables. The vector treatment of Newton's Second Law is given in Section 4.1, where we model pendulum motion.

All the basic facts about constant-coefficient second-order linear ODEs appear in the pages of this chapter—and more besides. Our approach, for the most part, is fairly standard, but here and there we have added some material that gives some deeper insight into some corners that are often overlooked. The only model in the chapter appears in the first section [we like to start off our chapters with a model whenever possible]. In addition to the standard linear spring, we also develop hard and soft springs, and aging springs. It is no more trouble to model all these types of springs in one fell swoop, so that is what we do. The hard and soft spring ODEs are nonlinear and provide us with a nice opportunity to compare solutions of nonlinear ODEs and their linearizations (see especially Fig. 3.1.3). We take up the aging spring model again in Chapter 11, where we show how to solve it using series methods.

The geometry of solutions of second-order ODEs is a bit more complicated than for first-order ODEs. Orbits and time state curves are introduced in Section 3.2 along with several examples. Section 3.2 also extends the Fundamental Theorem to a second-order initial value problem. This theorem gives conditions that ensure that an IVP has exactly one solution and also gives the extension and sensitivity properties of that solution. Sections 3.3–3.6 introduce constant-coefficient linear second-order ODEs with and without initial data. The basic tool for finding solution formulas is the use of the operator $P(D) = D^2 + aD + b$ where a and b are real numbers. So we spend some time talking about this operator. The emphasis of this chapter is on linear ODEs, because the solution set of a second-order linear ODE can be characterized in simple terms. In Sections 3.3 and 3.4 we show how to find all solutions if the second-order linear ODE is undriven and has constant coefficients. In Section 3.5 we look at undriven linear ODEs with periodic solutions and the difficulty in tracking such solutions numerically.

The Method of Undetermined Coefficients for linear homogeneous second-order ODEs with constant coefficients and polynomial-exponential driving terms appears in examples in Section 3.3 and is extensively developed in Section 3.6.

In Section 3.7 we present the general theory of linear ODEs using Wronskians and basic solution sets. Variation of parameters "works" for every second-order linear ODE, $y'' + a(t)y' + b(t)y = f(t)$, and we present the method in a lengthy problem at the end of Section 3.7.

3.1 Springs: Linear and Nonlinear Models

Comments

In this section we model various kinds of springs: Hooke's Law springs where the spring force is proportional to displacement, hard and soft springs where the force has extra terms in the cube of the displacement, and the aging spring, which like the Hooke's Law case leads to a linear ODE but with a time-varying coefficient of proportionality. The regions of validity of the model ODEs are discussed. Of particular importance, we show how the nonlinear spring model ODEs can be linearized near an equilibrium point.

This section is mostly introductory, presenting not much theory, but lots of examples and comments about the spring model. Note that force models (typically) lead to second-order ODEs in a position variable and that second-order ODEs require two state variables, usually position and velocity. Second-order ODEs play a starring role in an ODE course because of the force laws. We often refer to the spring models later in this chapter and in subsequent chapters.

---

1. **(a).** The motion satisfies equation (8) of the text, $mz'' + cz' + kz = f(t)$. Since there is no damping or driving force, $c = 0$ and $f = 0$, and the equation becomes $z'' + (k/m)z = 0$; so we need to find k/m. Since $kh = mg$, $k/m = g/h$, where $g = 384$ in/sec^2 [see Example 3.1.1], $k/m = g/h = (384 \text{ in/sec}^2)/(24 \text{ in}) = 16 \text{ sec}^{-2}$. So the equation of motion is $z'' + 16z = 0$.

(c). See Fig. 1(c) for the plot in tz-space (top) and in tz'-space (bottom). The motion is periodic, and all nonconstant solutions seem to have period $T \approx 1.6$ sec (reading from the tz-graph at top right).

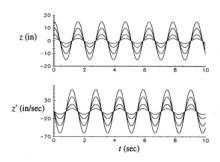

Problem 1(c).

3. Whenever we interpret the results of our model in terms of the system being modeled, we must be careful to restrict our attention to the regions where our model is valid. For springs, the model is valid only when the displacement is within the elastic limits of the spring. That is, the model is only valid for displacements yielding a restoring spring force, positive for $y < 0$ and negative for $y > 0$. For the soft springs of **(b)** and **(c)**, the elastic limit is no greater than $(10)^{1/2} \approx 3.16$, since for $y > (10)^{1/2}$, $S(y)$ will have the wrong sign for a restoring force. A cyclical orbit [as in **(a)** and **(b)**] indicates that after a certain period of time the system returns to its original configuration, and since time does not appear explicitly in the equations of motion (i.e., the ODEs are autonomous) the system repeats its motion periodically. Generally, the period will be different for different orbits, as a component plot will show. In **(c)** the motion is damped, so energy is dissipated and the amplitude decays with time. In every case $m = 1$ and the corresponding system is $y' = v$, $v' = S(y) - cy'$.

(a). Here we have $y' = v$, $v' = -0.2y - 0.02y^3$, $0 \le t \le 25$. See Figs. 3(a). The equilibrium point at the origin of Fig. 3(a), Graph 1 and the corresponding straight line solution curve in Fig. 3(a), Graph 2 correspond to the spring at rest at $y = 0$ in its position of static equilibrium. The other three solutions are periodic and correspond to the periodic oscillations of the undamped spring. The periods of the three solutions corresponding to $y_0 = 0$, $v_0 = 1, 3, 9$ are approximately 12.3, 8.7, and 5.5, respectively; the period appears to *decrease* as the amplitude increases. Apparently the stiffness of the spring is such that the spring oscillates faster and with greater amplitude if it is given a greater initial velocity.

(c). Here, $y' = v$, $v' = -y + 0.1y^3 - 0.1v$, $0 \le t \le 40$. The solution with initial point $y_0 = 0$, $v_0 = 2.44$ is just close enough to the rest point of static equilibrium $y = 0$, $v = 0$ [see Fig. 3(c), Graphs 1, 2], that the damping forces the motion of the spring into decaying oscillations around the equilibrium. The solution with initial data $y_0 = 0$, $v_0 = 2.46$ has just that extra amount of initial velocity (and so, energy) that it appears to move beyond the limits of validity of the model. See the orbit in Fig. 3(c), Graph 1 and the solution curve in Fig. 3(c), Graph 2 that "escape" to infinity as t increases.

Problem 3(a), Graph 1.

Problem 3(a), Graph 2.

Problem 3(c), Graph 1.

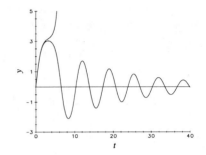

Problem 3(c), Graph 2.

5. We will try to account for gravity by simply changing variables to shift the equilibrium position of the spring from h in the old variable y to 0 in the new variable z as done in equation (8) in the text. Substituting $y = z - h$ in the hard/soft spring models, $ay'' + cy' + ky \pm jy^3 = -mg + f(t)$, with $S(y)$ given by (3) or (4) yields $mz'' + cz' + kz \pm j(z^3 - 3z^2h + 3zh^2) = f(t)$, where we have used the fact that the equilibrium displacement h is the value of y at which the gravitational force is exactly balanced by the spring force, so h satisfies $k(-h) \pm j(-h)^3 = -mg$. Since our transformed ODE for motion about the equilibrium changes by a nonzero term $\pm j(-3z^2h + 3zh^2) = \pm 3zhj(-z + h)$ when gravity is introduced, gravity *does* affect the motion of hard and soft springs more than just changing the position of static equilibrium, and it does so by changing the form of the ODE of the motion near the point of static equilibrium. In the Hooke's Law spring, this substitution does not change the equation of motion since $j = 0$ under Hooke's Law, so the Hooke's Law spring's motion is not affected by gravity.

7. Group project.

3.2 Second Order ODEs and Their Properties

Comments

Whenever we introduce a new class of ODEs (second-order ODEs in this chapter), we need to assure ourselves that IVPs have unique solutions. As usual, this requires that the functions in the ODE satisfy certain continuity and smoothness conditions. That is the reason Theorem 3.2.1 is given here. In addition, we want to know whether the solution of an IVP responds in a continuous fashion to changes in the initial data—and this is where the Continuity [or Sensitivity] part of Theorem 3.2.1 comes in. We also define and name the various graphs corresponding to a solution of an IVP. If the ODE has the form $y'' = F(t, y, y')$ and $y = y(t)$ is a solution, we discuss and illustrate solution curves [y vs. t], velocity curves [y' vs. t], orbits [y' vs. y], and time-state curves [the parametrically-defined curve $t = t$, $y = y(t)$, $y' = y'(t)$ in tyy'-space]. It takes a while to absorb all of this, and so we have given lots of examples with pictures. Each graph gives its own kind of information about the behavior of a solution; if we don't have a solution formula, we need all the help we can get to determine the behavior of a solution. Group problem 12 should be noted because it gives a way to trick your solver so that it will plot a time-state curve and its orbit on the same set of axes. As in the group project in Section 3.1, the method involves adding extra ODEs.

---

1.   To verify that each IVP satisfies Theorem 3.2.1, we must write the ODE in the form $y'' = F(t, y, y')$ and show that F, $\partial F / \partial y$, and $\partial F / \partial y'$ are continuous throughout tyy'-space. The ty-solution curves and ty'-velocity curves are plotted in Graph 1, then the orbits in the yy'-plane in Graph 2, and finally the time-state curves (t is the vertical axis) in tyy'-space in Graph 3.

(a). The ODE can be rewritten as $y'' = F(t, y, y') = -y$. The function $F = -y$ is continuous throughout tyy'-space as are $\partial F / \partial y = -1$, and $\partial F / \partial y' = 0$, so Theorem 3.2.1 is satisfied. The solutions $y = y(t)$ and velocities $v = y'(t)$ of the IVPs $y'' + y = 0$, $y(0) = 0.5$, 1.0, 1.5, $y'(0) = 0$ seem to be periodic with a common period T a little larger than 3. There is a single equilibrium solution, $y = 0$ for all t, whose graph is a straight line. The nonconstant orbits seem to be simple closed curves (which again suggests that the solutions are periodic). The equilibrium solution corresponds to a point orbit at the origin. The time-state curves are helices, while the equilibrium time-state curve is a vertical line. See Figs. 1(a).

Problem 1(a), Graph 1.

Problem 1(a), Graph 2.

Problem 1(a), Graph 3.

(c). The ODE can be rewritten as $y'' = F(t, y, y') = 0.1y' - 4y$. The function $F = 0.1y' - 4y$ is continuous throughout tyy'-space as are $\partial F/\partial y = -4$ and $\partial F/\partial y' = 0.1$, so Theorem 3.2.1 is satisfied. Here the IVPs are $y'' - 0.1y' + 4y = 0$, $y(0) = 0, 1, 2$, $y'(0) = 0$, but we solve *backward* from $t = 0$ to $t = -100$. There is a single equilibrium solution, $y = 0$, all t. See Figs. 1(c).

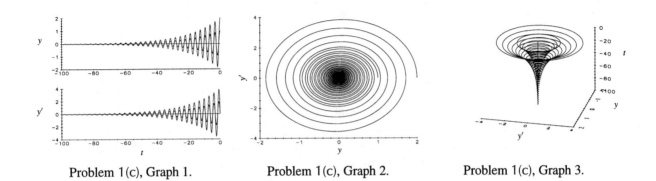

Problem 1(c), Graph 1.   Problem 1(c), Graph 2.   Problem 1(c), Graph 3.

(e). The ODE can be rewritten as $y'' = F(t, y, y') = -(0.1y^2 - 0.08)y' - y^3$. The function $F = -(0.1y^2 - 0.08)y' - y^3$ is continuous, as are $\partial F/\partial y = -0.2yy' - 3y^2$ and $\partial F/\partial y' = -0.08$, so Theorem 3.2.1 is satisfied. The solution of the nonlinear IVP $y'' + (0.1y^2 - 0.08)y' + y^3 = 0$, $y(0) = 0$, $y'(0) = 0.5$ seems to oscillate with a slowly growing amplitude [Fig. 1(e), Graph 1]. The orbit [Fig. 1(e), Graph 2] suggests that as $t \to +\infty$ the orbit spirals to the orbit of a slightly tilted, periodic "racetrack" orbit. Turning back to the solution curve in Fig. 1(e), Graph 1, we see from the solution curve for large t that the period of the racetrack orbit is about 4.4. The time-state curve is shown in Fig. 1(e), Graph 3.

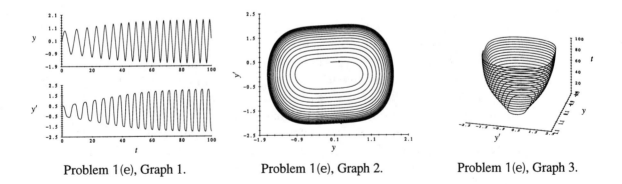

Problem 1(e), Graph 1.   Problem 1(e), Graph 2.   Problem 1(e), Graph 3.

3. See Fig. 3 for graphs of several solution curves of $t(y''y + (y')^2) + y'y = 1$ with initial points $y(1) = 1$, $|y'(1)| = |v_0| \leq 10$.

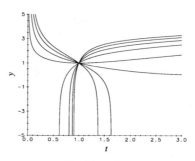

Problem 3.

5. To solve the IVP $y'' = 2y'y$, $y(0) = 1$, $y'(0) = 1$ we use the hint to see that $y'' = (y^2)'$. Integrating and applying the initial conditions, we see that $y' = y^2$, so $y'/y^2 = 1$. Integrating again and applying the initial conditions, we obtain $-1/y + 1 = t$. Solving for y, we have $y = 1/(1 - t)$. The largest t-interval for this IVP is $t < 1$. See Figs. 5, Graph 1 and Graph 2 for the solution graph and orbit, respectively.

Problem 5, Graph 1.

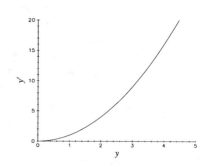

Problem 5, Graph 2.

7. Look for solutions to the IVP $t^2 y'' - 2ty' + 2y = 0$, $y(0) = 0$, $y'(0) = 0$ in the form $y = Ct^\alpha$. Substituting this expression into the system, we obtain $C\alpha(\alpha - 1)t^\alpha - 2C\alpha t^\alpha + 2Ct^\alpha = Ct^\alpha[\alpha(\alpha - 1) - 2\alpha + 2] = 0$. For a nontrivial solution, $C \neq 0$, and so $\alpha(\alpha - 1) - 2\alpha + 2 = \alpha^2 - 3\alpha + 2$ must be zero. So, $\alpha = 1$ or $\alpha = 2$. This means that the original ODE has solutions $y = C_1 t + C_2 t^2$ where C_1 and C_2 are arbitrary constants. In order to satisfy the initial conditions, $y(0) = 0$, $y'(0) = 0$, we must have $C_1 = 0$. The IVP has infinitely many solutions $y = C_2 t^2$, where C_2 is an arbitrary constant. This conclusion does not contradict Theorem 3.2.1 because $F(t, y, y') = (2ty' - 2y)/t^2$ is not continuous with respect to t at $t = 0$.

9. If we set $v = y'(t)$, $vdv/dy = y''(t)$ [see Section 1.7], then the ODE, $y'' = -ky - jy^3$ becomes the first-order ODE

$$v\frac{dv}{dy} = -ky - jy^3$$

which is separable. Separating the variables,

$$v \, dv = (-ky - jy^3) \, dy$$

Integrating each side, we have

$$\frac{1}{2}v^2 = -\frac{1}{2}ky^2 - \frac{j}{4}y^4 + C_1$$

$$v^2 = (y')^2 = C - ky^2 - jy^4/2$$

where $C = 2C_1$ is a positive constant. For each positive value of C, the graph is an oval-shaped curve in the yv-plane which cuts the y-axis at the real roots of $C - ky^2 - jy^4/2$. These real roots can be found by using the quadratic formula for y^2, and then taking the square root of the positive roots. In fact, we have that the real roots are given by

$$y = \pm \left[-\frac{k}{j} + \left(\frac{k^2}{j^2} + \frac{2C}{j} \right)^{1/2} \right]^{1/2}$$

11. According to Theorem 3.2.1, if F, $\partial F/\partial y$, and $\partial F/\partial y'$ are continuous for all t, y, y', then the IVP, $y'' = F(t, y, y')$, $y(t_0) = y_0$, $y'(t_0) = v_0$ has exactly one solution.

(a). The functions $y_1 = \sin t$, and $y_2 = t$ both have value 0 at $t_0 = 0$, and the derivatives at $t_0 = 0$ have common value 1. So if these two functions were both solutions of $y'' = F(t, y, y')$, then there would be *two* solutions satisfying the conditions $y(0) = 0$, $y'(0) = 1$, and this would contradict Theorem 3.2.1. So $y_1(t)$ and $y_2(t)$ cannot both be solutions of $y'' = F(t, y, y')$.

3.3   Undriven Constant Coefficient Linear ODEs, I

Comments

Section 3.3 lays the foundation for a study of undriven second-order linear ODEs with constant coefficients and the corresponding IVPs. Only the case where the characteristic roots are real is discussed. We take the linear operator approach, but gently. The next section treats undriven ODEs with constant coefficients whose characteristic roots are complex numbers.

---

1. Although it is not necessary to include the step using operator notation, we put it in to help you grow accustomed to its use and to help you see how the characteristic polynomial $P(r)$ relates to the polynomial operator $P(D)$.

(a). The ODE $y'' = D^2[y] = 0$ has the characteristic polynomial r^2 which has a double root, $r = 0$. Using Theorem 3.3.1, the general solution is $y(t) = C_1 e^{0t} + C_2 t e^{0t} = C_1 + C_2 t$, where C_1 and C_2 are arbitrary reals. Notice that you could also integrate both sides twice to get the same result.

(c). The ODE $y'' - 4y' + 4y = (D^2 - 4D + 4)[y] = 0$ has the characteristic polynomial $r^2 - 4r + 4$ which has a double root, $r = 2$. Using Theorem 3.3.1, the general solution is $y(t) = C_1 e^{2t} + C_2 t e^{2t}$, where C_1 and C_2 are arbitrary reals.

(e). The ODE $5y'' - 10y' = (5D^2 - 10D)[y] = 0$ has the characteristic polynomial $5r^2 - 10r$ which has roots $r_1 = 0$ and $r_2 = 2$. So the general solution is $y(t) = C_1 + C_2 e^{2t}$.

(g). The ODE $y'' - 6y' + 9y = (D^2 - 6D + 9)[y] = 0$ has the characteristic polynomial $r^2 - 6r + 9$ which has a double root, $r = 3$. So the general solution is $y(t) = C_1 e^{3t} + C_2 t e^{3t}$.

(i). The ODE $y'' + 2y' + y = (D^2 + 2D + 1)[y] = 0$ has the characteristic polynomial $r^2 + 2r + 1$ which has a double root, $r = -1$. So the general solution is $y(t) = C_1 e^{-t} + C_2 t e^{-t}$.

(k). The ODE $y'' - 2y' + y = (D^2 - 2D + 1)[y] = 0$ has the characteristic polynomial $r^2 - 2r + 1$ which has a double root, $r = 1$. So the general solution is $y(t) = C_1 e^t + C_2 t e^t$.

3. Theorem 3.3.1 shows us that all possible solutions of $y'' + ay' + by = (D^2 + aD + b)[y] = P(D)[y] = 0$ are of the form $C_1 e^{r_1 t} + C_2 e^{r_2 t}$ or $C_1 e^{r_1 t} + C_2 t e^{r_1 t}$ where $P(r) = r^2 + ar + b = (r - r_1)(r - r_2)$. Solving for these cases is a matter of matching terms in the given solution with the appropriate roots. The values of C_1 and C_2 don't affect our choice of differential equation (these values are fixed when initial conditions are chosen).

(a). In the case of $e^t - e^{-t}$ we see that the roots are $r_1 = 1$ and $r_2 = -1$ (we don't really care about C_1 or C_2). So the characteristic polynomial $P(r)$ must have as roots 1 and -1. So $P(r) = (r - r_1)(r - r_2) = (r - 1)(r + 1) = r^2 - 1$. So we know that $P(D) = D^2 - 1$ and $P(D)[y] = (D^2 - 1)[y] = y'' - y$. Our original ODE is $y'' - y = 0$.

(c). In the case of $e^t + e^{-2t}$ we see that the roots are $r_1 = 1$ and $r_2 = -2$. So $P(r) = (r - 1)(r + 2) = r^2 + r - 2$. So $P(D) = D^2 + D - 2$ and $P(D)[y] = (D^2 + D - 2)[y] = y'' + y' - 2y$. So the original ODE is $y'' + y' - 2y = 0$.

(e). In the case of $e^{2t} + 10000e^{3t}$ we see that the roots are $r_1 = 2$ and $r_2 = 3$. So $P(r) = (r - 2)(r - 3) = r^2 - 5r + 6$. So $P(D) = D^2 - 5D + 6$ and $P(D)[y] = (D^2 - 5D + 6)[y] = y'' - 5y' + 6y$. So the original ODE is $y'' - 5y' + 6y = 0$.

(g). In the case of $e^{\pi t} - 3$ we see that the roots are $r_1 = \pi$ and $r_2 = 0$. So $P(r) = (r - \pi)r = r^2 - r\pi$ giving us $P(D) = D^2 - D\pi$ and $P(D)[y] = (D^2 - D\pi)[y] = y'' - \pi y'$. So the original ODE is $y'' - \pi y' = 0$.

(i). In the case of $t + 2$ we see that it can be written in the form $2e^{0t} + te^{0t}$ where both roots are zero. So $P(r) = r^2$ giving $P(D) = D^2$ and $P(D)[y] = D^2[y] = y''$. So the original ODE must be $y'' = 0$.

5. **(a).** $(D - 2)[e^{-t}] = D[e^{-t}] - 2e^{-t} = -e^{-t} - 2e^{-t} = -3e^{-t}$.

(c). $(D^2 + D - 6)[e^{-t}] = D^2[e^{-t}] + D[e^{-t}] - 6e^{-t} = e^{-t} - e^{-t} - 6e^{-t} = -6e^{-t}$.

(e). Using the fact from part **(a)** that $(D - 2)[e^{-t}] = -3e^{-t}$,

$$(D + 3)(D - 2)[e^{-t}] = (D + 3)[-3e^{-t}]$$
$$= 3e^{-t} - 9e^{-t}$$
$$= -6e^{-t}$$

which equals the results in parts **(c)** and **(d)** since $D^2 + D - 6 = (D+3)(D-2) = (D-2)(D+3)$.

7.   **(a).** We will guess that $y(t) = Ae^{-t}$ for some value of A. Then

$$(D-2)[Ae^{-t}] = -Ae^{-t} - 2Ae^{-t}$$

$$= -3Ae^{-t}$$

$$= e^{-t}$$

is accomplished by setting $A = -1/3$, so $y(t) = (-1/3)e^{-t}$.

(c). We will guess that $y(t) = A\cos t + B\sin t + C$ for some choice of A, B, C. Then

$$(D-2)[y(t)] = -A\sin t + B\cos t - 2A\cos t - 2B\sin t - 2C$$

$$= (-A - 2B)\sin t + (B - 2A)\cos t - 2C$$

$$= \sin t + 4$$

So $A = -1/5$, $B = -2/5$, $C = -2$, and $y(t) = -1/5\cos t - 2/5\sin t - 2$.

9.   If y solves the ODE $(D-1)(D+2)(D-3)[y] = 0$, then $v = (D+2)(D-3)[y]$ solves the first-order ODE $(D-1)[v] = 0$. Solving this ODE we obtain $v = C_1 e^t$, where C_1 is an arbitrary constant. Now if y solves $(D+2)(D-3)[y] = v$, then $w = (D-3)[y]$ solves the ODE $(D+2)[w] = v = C_1 e^t$. Now let's solve $w' + 2w = C_1 e^t$. Multiplying through by the integrating factor e^{2t}, this ODE becomes $(we^{2t})' = C_1 e^{3t}$. Integrating, we have $we^{2t} = (C_1/3)e^{3t} + C_2$, where C_2 is an arbitrary real. So we may just as well write $w = C_1 e^t + C_2 e^{-2t}$ where we have replaced $C_1/3$ by C_1 (which is just as arbitrary as $C_1/3$). Finally, we solve the ODE $(D-3)[y] = w = C_1 e^t + C_2 e^{-2t}$ for y. Write the ODE as $y' - 3y = C_1 e^t + C_2 e^{-2t}$ and multiply through by the integrating factor e^{-3t} to get $(ye^{-3t})' = C_1 e^{-2t} + C_2 e^{-5t}$. Integration gives us that $ye^{-3t} = -(C_1/2)e^{-2t} - (C_2/5)e^{-5t} + C_3$, where C_3 is an arbitrary constant. Solving for y we get the general solution of the original ODE,

$$y = C_1 e^t + C_2 e^{-2t} + C_3 e^{3t}$$

where we have replaced $-C_1/2$ by C_1 and $-C_2/5$ by C_2 (because C_1 and C_2 are just as arbitrary as $-C_1/2$ and $C_2/5$).

Notice that this result suggests that Theorem 3.3.1 can be extended to linear ODEs of any order.

There is another approach to finding a general solution formula for the constant-coefficient ODE $P(D)[y] = 0$, which will be especially useful when the coefficients of $P(D)$ are not constants. Let's see how this approach goes.

Background Material: Another General Solution Formula Approach

From the General Solution Theorem 3.3.1 we saw that if the solution pair $\{y_1, y_2\}$ is cleverly chosen, then all solutions of the ODE $P(D)[y] = 0$ are given by taking all linear combinations of these solutions. It turns out that there are infinitely many solution pairs

$\{y_1, y_2\}$ that will also do the job. To see this, suppose that $\{y_1(t), y_2(t)\}$ is a pair of solutions of the ODE $P(D)[y] = 0$. If c_1 and c_2 are any constants, then from the Closure Property we know that the linear combination $c_1 y_1 + c_2 y_2$ is also a solution of that ODE.

Now if $z(t)$ is any solution at all, then we claim that there are unique constants $\tilde{c}_1$ and $\tilde{c}_2$ such that $z = \tilde{c}_1 y_1 + \tilde{c}_2 y_2$, provided that the determinant

$$W(t) = \det \begin{bmatrix} y_1(t) & y_2(t) \\ y_1'(t) & y_2'(t) \end{bmatrix} = y_1(t) y_2'(t) - y_1'(t) y_2(t) \tag{i}$$

is nonzero at a point t_0 (any point t_0 will do). The function $W(t)$ is called the *Wronskian* of the solution pair $\{y_1, y_2\}$ and is sometimes written as $W[y_1, y_2](t)$ to emphasize the dependence on the solution pair.

Suppose now that t_0 is a point where $W(t_0) \neq 0$. Let's look at the solvability of the algebraic equations

$$\begin{aligned} c_1 y_1(t_0) + c_2 y_2(t_0) &= z(t_0) \\ c_1 y_1'(t_0) + c_2 y_2'(t_0) &= z'(t_0) \end{aligned} \tag{ii}$$

for the constants c_1 and c_2. The determinant of this algebraic system is $W[y_1, y_2](t_0)$, the Wronskian of the solution pair $\{y_1, y_2\}$ evaluated at t_0. Since by assumption $W[y_1, y_2](t_0) \neq 0$, the algebraic system (ii) has a unique solution $\tilde{c}_1, \tilde{c}_2$. Now put

$$w(t) = \tilde{c}_1 y_1(t) + \tilde{c}_2 y_2(t) - z(t)$$

and notice that $w(t)$ solves IVP (10), and so $w(t)$ is the zero function. This implies that $z(t) = \tilde{c}_1 y_1(t) + \tilde{c}_2 y_2(t)$, and so $z(t)$ is indeed a linear combination of $y_1(t)$ and $y_2(t)$. It follows that

$$y = c_1 y_1 + c_2 y_2, \quad c_1, c_2 \text{ arbitrary constants}$$

is the general solution of $P(D)[y] = 0$.

So apparently any solution pair whose Wronskian is nonzero at some (any) point is very important. We need a name for such solution pairs:

❖ **Basic Solution Set**. The pair of solutions $\{y_1, y_2\}$ of the ODE $y'' + ay' + by = 0$ is called a *basic solution set* if there is a point t_0 such that the Wronskian $W[y_1, y_2](t_0) \neq 0$.

Actually, the Wronskian of a solution pair is either always zero or never zero.

So the general solution of $P(D)[y] = 0$ is given by $y = c_1 y_1 + c_2 y_2$, for arbitrary constants c_1 and c_2, if $\{y_1, y_2\}$ is a basic solution set. As we show next, basic solution sets are easy to construct if the roots of the characteristic polynomial $r^2 + ar + b$ are real.

Let's examine the solution pairs that appear in the general solution to the ODE $y'' + ay' + by = 0$. If the roots r_1 and r_2 of the characteristic polynomial satisfy $r_1 \neq r_2$, we have the solution pair $y_1 = e^{r_1 t}$ and $y_2 = e^{r_2 t}$. The Wronskian

$$W[y_1, y_2] = e^{r_1 t}(e^{r_2 t})' - (e^{r_1 t})'(e^{r_2 t})$$

$$= (r_2 - r_1) e^{(r_1 + r_2)t}$$

is always nonzero, and so this solution pair is basic.

If $r_1 = r_2$, we have the solution pair $y_1 = e^{r_1 t}$ and $y_2 = te^{r_1 t}$. The Wronskian

$$W[y_1, y_2] = e^{r_1 t}(te^{r_1 t})' - (e^{r_1 t})'(te^{r_1 t})$$

$$= e^{r_1 t}(e^{r_1 t} + r_1 te^{r_1 t}) - (r_1 e^{r_1 t})(te^{r_1 t})$$

$$= e^{2r_1 t}$$

is never zero, and so this solution pair is also basic.

3.4 Undriven Constant Coefficient Linear ODEs, II

Comments

We characterize the solution set of the second-order, constant coefficient, undriven, linear ODE, no matter whether the characteristic roots are real or complex. The differentiation operator is generalized to act on complex-valued functions of a real variable. Example 3.4.3 shows how to get real solutions of a real ODE from complex solutions. We like to do a lot of graphing of solution curves, especially with $y(t_0)$ fixed, but $y'(t_0)$ varying, as initial data. This highlights once more that solution curves of normalized second-order ODEs require two items of initial data to determine a unique solution, that solution curves of a second-order ODE can intersect with impunity, and that one can imagine each point (t_0, y_0) in the ty-plane as a source of solution curves emerging from the point in all directions (except vertical). Problems 5 and 6 are particularly important in the later applications because they give the conditions for stability or asymptotic stability (or as engineers might say in the latter case, all solutions are "transients").

---

1. The general solution of $P(D)[y] = 0$ is given in each case.

   (a). Since the characteristic polynomial of $P(D) = D^2 - D - 2$ is $r^2 - r - 2$ whose roots are -1 and 2, we have $y = c_1 e^{-t} + c_2 e^{2t}$, where c_1 and c_2 are arbitrary real constants.

   (c). The characteristic polynomial of $D^2 - 4D + 4$ is $r^2 - 4r + 4$ which has a double root $r = 2$. The general solution is $y = (c_1 + c_2 t)e^{2t}$, where c_1 and c_2 are arbitrary real constants.

3. In each part, solution curves are plotted in Graph 1, with the corresponding orbits plotted in Graph 2; c_1 and c_2 are arbitrary real constants.

   (a). The characteristic polynomial for $y'' + y' = (D^2 + D)[y] = 0$ is $r^2 + r$, which factors to $(r + 1)r$. Thus, solutions of the ODE are given by e^{-t}, $e^{0 \cdot t}$, and the linear combinations of these functions. That is, the real-valued solutions are given by $y = c_1 e^{-t} + c_2$. See Figs. 3(a). From the figures and the solution formulas we see that as $t \to +\infty$, $y(t) \to c_2$, while (if $c_1 \neq 0$), as $t \to -\infty$, $y(t) \to \pm\infty$ depending on the sign of c_1. The orbits are slanted rays that approach the y-axis in the yy'-plane as $t \to +\infty$.

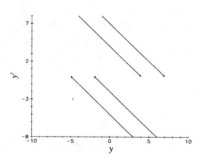

Problem 3(a), Graph 1.   Problem 3(a), Graph 2.

(c). The characteristic polynomial of the ODE $y'' + 3y' + 2y = (D^2 + 3D + 2)[y] = 0$ is $r^2 + 3r + 2$ which factors to $(r + 1)(r + 2)$. Solutions of the ODE are given by e^{-t}, e^{-2t}, and the linear combinations of these functions. So the real-valued solutions are given by $y = c_1 e^{-t} + c_2 e^{-2t}$. See Fig. 3(c), Graph 1. From the formula and the pictures as $t \to +\infty$ solutions decay to 0, but as $t \to -\infty$, solutions become unbounded. The orbits tend to the origin of the yy'-plane as $t \to +\infty$ [Fig. 3(c), Graph 2].

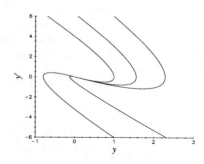

Problem 3(c), Graph 1.   Problem 3(c), Graph 2.

(e). The characteristic polynomial of the ODE $y'' - y/9 = (D^2 - 1/9)[y] = 0$ is $r^2 - 1/9$ which factors to $(r + 1/3)(r - 1/3)$. Solutions of the ODE are given by $e^{-t/3}$, $e^{t/3}$, and the linear combinations of these functions. The real-valued solutions are given by $y = c_1 e^{-t/3} + c_2 e^{t/3}$. See Fig. 3(e), Graph 1. The solution curves become unbounded as $t \to \pm\infty$ if c_1 and c_2 are nonzero. The time span used here is fairly short, so we only see arcs of orbits (in some cases) [Fig. 3(e), Graph 2] that would move out of the rectangle if the time span, both forward and backward, were longer. In Fig. 3(e), Graph 3 we have enlarged the xy-rectangle and lengthened the time span, so we can really see what is going on.

Problem 3(e), Graph 1.   Problem 3(e), Graph 2.   Problem 3(e), Graph 3.

5. **(a).** If $a > 0$ and $b > 0$, then the roots of the characteristic polynomial $r^2 + ar + b$ are $-a/2 \pm (a^2/4 - b)^{1/2}$. If $(a/2)^2 > b$, then both roots r_1 and r_2 are negative since $((a/2)^2 - b)^{1/2} < a/2$. In this case, the general solution $y = c_1 e^{r_1 t} + c_2 e^{r_2 t} \to 0$ as $t \to +\infty$. If $b = (a/2)^2$, then there is a double root $-a/2$ and the general solution is $y = c_1 e^{r_1 t} + c_2 t e^{r_1 t}$ and again $y \to 0$ as $t \to +\infty$ since r_1 is negative and $e^{r_1 t} \to 0$ faster than $t \to +\infty$ (apply L'Hopital's Rule to $t/e^{-r_1 t}$). If $b > (a/2)^2$, then the roots $\alpha \pm i\beta$ are imaginary with real part $-a/2 = \alpha$, which is negative. The general solution $y = e^{\alpha t}(c_1 \cos \beta t + c_2 \sin \beta t) \to 0$ as $t \to +\infty$ because $e^{\alpha t} \to 0$, while $c_1 \cos \beta t + c_2 \sin \beta t$ oscillates but remains bounded.

7. **(a).** Using the quadratic formula, we find that the roots of the characteristic polynomial $r^2 + ir + 2$ are $-2i$ and i. The general solution of the ODE $y'' + iy' + 2y = 0$ is $y = k_1 e^{-2it} + k_2 e^{it}$, where k_1 and k_2 are any complex constants.

(c). Let z be a root of $r^2 + i$. Then $z = r_0 e^{i\theta}$ and $z^2 = r_0^2 e^{2i\theta} = -i = e^{(3\pi/2 + 2n\pi)i}$. Matching magnitudes and angles, we have that $r_0 = 1$ and $2\theta = 3\pi/2 + 2n\pi$. The only two distinct angles correspond to $n = 0, 1$. So, $r_0 = 1$, $\theta_0 = 3\pi/4$, $\theta_1 = 7\pi/4$, and the two roots of $r^2 + i$ are $e^{3\pi i/4} = (-1 + i)/\sqrt{2}$ and $e^{7\pi i/4} = (1 - i)/\sqrt{2}$. The general solution is $y = k_1 e^{(-1+i)t/\sqrt{2}} + k_2 e^{(1-i)t/\sqrt{2}}$, where k_1 and k_2 are any complex constants.

Background Material: Another Approach to Finding Solution Formulas

In the background material at the end of Section 3.3 of this Student Resource Manual, we showed that if we could find a basic solution pair $\{y_1, y_2\}$ for the ODE $y'' + ay' + by = 0$, then the general solution could be written as

$$y = c_1 y_1 + c_2 y_2$$

where c_1 and c_2 were arbitrary constants. In that section we assumed that the solutions y_1 and y_2 were real-valued, so the arbitrary constants were reals. It will come as no surprise that the notion of a basic solution pair extends to complex-valued solutions, but with the difference that the constants c_1 and c_2 this time are arbitrary complex numbers. For example, in Theorem 3.4.2 note that if r_1 and r_2 are the characteristic roots, the ODE becomes

$$(D - r_1)(D - r_2)[y] = 0$$

so $y_1 = e^{r_1 t}$ and $y_2 = e^{r_2 t}$ are solutions. If $r_1 \neq r_2$ (r_1 and r_2 need not be real), then the Wronskian $W[e^{r_1 t}, e^{r_2 t}] = (r_2 - r_1)e^{(r_1 + r_2)t} \neq 0$, so $\{e^{r_1 t}, e^{r_2 t}\}$ is a basic solution pair. If $r_1 = r_2$, notice that $(D - r_1)^2[te^{r_1 t}] = 0$, so we have the solution pair $y_1 = e^{r_1 t}$, $y_2 = te^{r_1 t}$. Since the Wronskian $W[e^{r_1 t}, te^{r_1 t}] = e^{2r_1 t} \neq 0$, it follows that $\{e^{r_1 t}, te^{r_1 t}\}$ is a basic solution pair in this case.

The case $r_1 = \bar{r}_2 = \alpha + i\beta$ in Theorem 3.4.1 merits a special comment. In this case we have the two complex-valued solutions $e^{(\alpha+i\beta)t}$ and $e^{(\alpha-i\beta)t}$. Now from text formula (8),

$$e^{(\alpha+i\beta)t} = e^{\alpha t} \cos \beta t + ie^{\alpha t} \sin \beta t$$

Since $P(D) = D^2 + aD + b$ is a linear operator we see that

$$0 = P(D)[e^{\alpha t} \cos \beta t + ie^{\alpha t} \sin \beta t]$$
$$= P(D)[e^{\alpha t} \cos \beta t] + iP(D)[e^{\alpha t} \sin \beta t]$$

So *if the coefficients a and b are real* we conclude that

$$P(D)[e^{\alpha t} \cos \beta t] = 0, \quad \text{and} \quad P(D)[e^{\alpha t} \sin \beta t] = 0$$

so $y_1 = e^{\alpha t} \cos \beta t$ and $y_2 = e^{\alpha t} \sin \beta t$ comprise a real-valued solution pair. Since the Wronskian

$$W[e^{\alpha t} \cos \beta t, e^{\alpha t} \sin \beta t] = \beta e^{2\alpha t} \neq 0$$

it follows that $\{e^{\alpha t} \cos \beta t, e^{\alpha t} \sin \beta t\}$ is a basic solution pair for real-valued solutions.

In practice we prefer to use the basic solution pair approach, mostly because it generalizes to linear undriven ODEs with *nonconstant* coefficients (we show this in Section 3.7), and to higher order linear ODEs as well.

From Theorem 3.4.1, solutions of the constant-coefficient ODE $P(D)[y] = 0$ have oscillatory long-term behavior only if the roots of the characteristic equation $P(r) = 0$ are not real.

3.5 Periodic Solutions and Simple Harmonic Motion

Comments

Periodic solutions play an important role in the study of ODEs, so we thought we should give them a section of their own. Readers probably have seen periodic functions before, but they may have forgotten the terminology associated with them. It is especially important to read the discussion of when the sum of two periodic functions is periodic (and how to find the period of the sum). The simple harmonic motion examples (Examples 3.5.1 and 3.5.3) should be read carefully. Example 3.5.2 brings up the important point that when tracking a periodic solution (or an oscillating solution for that matter) with a numerical solver, decreasing the number of plotted points per unit time causes an interesting modulation effect on the output (aliasing). Not much is made of this phenomenon other than to say that when using a numerical solver it's important to try various numbers of plotted points before accepting the results. Incidentally, the aliasing effect is the reason stage-coach wheels in the movies often seem to rotate slowly or even rotate backwards. It all depends on the frame rate of the movie cameras.

1. **(a).** If $f(t)$ has (fundamental) period T, then T is the smallest positive number such that for any time t, $f(t+T) = f(t)$. Let $g(t) = f(\omega t)$ and suppose that $g(t)$ has (fundamental) period T_1; that is, T_1 is the smallest positive number such that $g(t+T_1) = g(t)$. Then $f(\omega(t+T_1)) = f(\omega t)$ by the definition of $g(t)$, so $f(\omega t + \omega T_1) = f(\omega t)$. But the smallest positive value of T_1 for which this holds satisfies $\omega T_1 = T$, since T is the period of f. So, $T_1 = T/\omega$ is the period of $g(t) = f(\omega t)$.

    (c). See Fig. 1(c) for the graphs of $\sin 2t$ (long dashes) and $\sin 5t$ (short dashes), with periods $T = \pi$ and $S = 2\pi/5$, respectively. Using the notation of part **(b)**, we have $m = 2$ and $n = 5$ so that $k = 2(\pi) = 5(2\pi/5) = 2\pi$. The solid curve is the graph of $\sin 2t + \sin 5t$, which does indeed have period 2π.

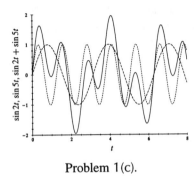

Problem 1(c).

3. To solve $(D^2 + 9)[y] = -3\cos 2t$, first note that the undriven solution is $C_1 \cos 3t + C_2 \sin 3t$. We need only find any particular solution to the driven equation. Guess a solution of the form $y_d = A\cos 2t + B\sin 2t$. Plugging this into the ODE, $(D^2 + 9)[A\cos 2t + B\sin 2t] = -4A\cos 2t + 9A\cos 2t - 4B\sin 2t + 9\sin 2t = -3\cos 2t$. We get $5A\cos 2t = -3\cos 2t$, $5B\sin 2t = 0$. So we must set $A = -3/5$, $B = 0$. The general solution is $y(t) = C_1 \cos 3t + C_2 \sin 3t - (3/5)\cos 2t$, which is periodic.

5. Group Project

3.6 Driven Constant Coefficient Linear ODEs

Comments

We try to do two things in this section: first, to get across the idea of the Method of Undetermined Coefficients, and second, to lead the reader through the algebraic intricacies of actually constructing a particular solution. A CAS would be helpful for the algebra. We have included many examples, because that is probably the only way to see how to carry out the undetermined coefficients approach. Although we do some numerical solving/plotting, the geometry of solution graphs is *not* the point of this section. The point is the construction of solution formulas. **Warning:** some problems involve a lot of crank work.

1. Use the Euler formulas, $\cos\theta = (e^{i\theta} + e^{-i\theta})/2$, $\sin\theta = i(e^{-i\theta} - e^{i\theta})/2$.

    (a). $\cos 2t - \sin t = (e^{2it} + e^{-2it})/2 + (e^{it} - e^{-it})i/2 = [i(e^{it} - e^{-it}) + (e^{2it} + e^{-2it})]/2$.

    (c). $t^2\sin 2t - (1+t)\cos^2 t = t^2\sin 2t - (1+t)(1 + \cos 2t)/2 = t^2(-e^{2it} + e^{-2it})i/2 - (1+t)[2 + (e^{2it} + e^{-2it})]/4 = -(1+t)/2 - (1+t+2it^2)e^{2it}/4 + (2it^2 - t - 1)e^{-2it}/4$.

    (e). $(1-t)e^{it}\cos 3t = (1-t)e^{it}(e^{3it} + e^{-3it})/2 = (1-t)(e^{4it} + e^{-2it})/2$.

3. In finding an annihilator a factor t^k suggests a root of multiplicity $k+1$ for the characteristic polynomial of the annihilator, while a complex exponent suggests a pair of complex conjugate roots.

    (a). We want to annihilate $t^2 e^{-it}$. The factor t^2 suggests a root of multiplicity 3; since $(D+i)[e^{-it}] = 0$, we guess and then verify that $(D+i)^3[t^2 e^{-it}] = 0$. Any polynomial operator with $(D+i)^3$ as a factor is an annihilator. So, an annihilator with real coefficients is $[(D+i)(D-i)]^3 = (D^2+1)^3$.

    (c). The function to be annihilated is $t + \sin t$. An annihilator with real coefficients for $t + \sin t$ is $D^2(D^2+1)$, since $D^2[t] = 0$ and $(D^2+1)[\sin t] = 0$.

5. **(a).** The IVP is $y'' - 4y = 2 - 8t$, $y(0) = 0$, $y'(0) = 5$. Note that $P(D) = (D-2)(D+2)$. A particular solution y_d of $(D-2)(D+2)[y] = 2 - 8t$ has the form

    $$y_d = A + Bt$$

    Since $(D-2)(D+2)[A + Bt] = -4(A + Bt) = 2 - 8t$, it follows that $A = -1/2$ and $B = 2$. So, a particular solution is $y_d = 2t - 1/2$. The characteristic polynomial $r^2 - 4$ has roots $r = \pm 2$, so the general solution of the undriven equation is $c_1 e^{-2t} + c_2 e^{2t}$. The general solution of the driven ODE is

    $$y = c_1 e^{-2t} + c_2 e^{2t} + 2t - 1/2$$

    Since $y(0) = 0$, $y'(0) = 5$, the solution of this IVP is

    $$y = -e^{-2t}/2 + e^{2t} + 2t - 1/2$$

    (c). The IVP is $y'' + y = 10e^{2t}$, $y(0) = 0$, $y'(0) = 0$. The ODE, $y'' + y = 10e^{2t}$, has $P(D) = D^2 + 1$. A particular solution has the form

    $$y_d = Ae^{2t}$$

    Insert y_d into the ODE to obtain

    $$5Ae^{2t} = 10e^{2t}$$

    Matching coefficients, we see that $A = 2$, so $y_d = 2e^{2t}$ is a particular solution. The characteristic polynomial $r^2 + 1$ has roots $\pm i$, so the general solution of the undriven equation is

$C_1 \cos t + C_2 \sin t$. The general solution of the driven ODE is

$$y = C_1 \cos t + C_2 \sin t + 2e^{2t}$$

Imposing the initial conditions, $y(0) = 0$, $y'(0) = 0$, we see that $0 = C_1 + 2$, $0 = C_2 + 4$, so the solution of the IVP is

$$y = -2\cos t - 4\sin t + 2e^{2t}$$

(e). The IVP is $y'' - 3y' + 2y = 8t^2 + 12e^{-t}$, $y(0) = 0$, $y'(0) = 2$. The ODE, $y'' - 3y' + 2y = 8t^2 + 12e^{-t}$, has $P(D) = (D-1)(D-2)$. A particular solution has the form

$$y_d = A + Bt + Ct^2 + De^{-t}$$

Inserting y_d into the ODE, we have

$$2A - 3B + 2C + (2B - 6C)t + 2Ct^2 + 6De^{-t} = 8t^2 + 12e^{-t}$$

Matching coefficients of like terms, we have

$$2A - 3B + 2C = 0, \quad 2B - 6C = 0, \quad 2C = 8, \quad 6D = 12$$

so $A = 14$, $B = 12$, $C = 4$, $D = 2$, and $y_d = 14 + 12t + 4t^2 + 2e^{-t}$ is a particular solution. The characteristic polynomial $r^2 - 3r + 2$ has roots 1, 2, so the general solution of the undriven equation is $C_1 e^t + C_2 e^{2t}$. The general solution of the driven ODE is

$$y = C_1 e^t + C_2 e^{2t} + 14 + 12t + 4t^2 + 2e^{-t}$$

Imposing the initial conditions, $y(0) = 0$, $y'(0) = 2$, we have

$$0 = C_1 + C_2 + 16, \quad 2 = C_1 + 2C_2 + 10$$

so $C_1 = -24$, $C_2 = 8$. The IVP has the solution

$$y = 8e^{2t} - 24e^t + 2e^{-t} + 4t^2 + 12t + 14$$

7. **(a).** First we find a particular solution of the ODE $(D^2 - D - 2)[y] = 2e^t$. Writing $P(D) = D^2 - D - 2$, we guess a particular solution of the form $y_d = Ae^t$. $P(D)[Ae^t] = -2Ae^t$, so we set $A = -1$. Since the roots of $P(r)$ are $r_1 = 2$, $r_2 = -1$ it follows from Theorem 3.4.1 that $y = C_1 e^{2t} + C_2 e^{-t}$, where C_1 and C_2 are arbitrary constants, is the general solution of $P(D)[y] = 0$. So $y = C_1 e^{2t} + C_2 e^{-t} - e^t$ is the general solution of $P(D)[y] = f(t)$. The arbitrary constants C_1 and C_2 in the general solution $y = C_1 e^{2t} + C_2 e^{-t} - e^t$ are determined by the initial data; $y(0) = 0 = C_1 + C_2 - 1$, $y'(0) = 1 = 2C_1 - C_2 - 1$. So, $C_1 = 1$, $C_2 = 0$, and the solution of the IVP is $y = e^{2t} - e^t$.

9. To find a solution for the equation $y'' + 25y = \sin \omega t$ where $\omega \neq 5$, first we find the undriven solution and then we find a particular solution. The general real-valued solution of the undriven equation $y'' + 25y = 0$ is $y = C_1 \cos 5t + C_2 \sin 5t$. To find a particular solution, first note that $\sin \omega t = \text{Im}[e^{i\omega t}]$. So we first solve $(D^2 + 25)[z] = e^{i\omega t}$ where $y = \text{Im}[z]$. Substituting $z = Ae^{i\omega t}$ into the ODE yields $A = 1/(25 - \omega^2)$. Setting $y = \text{Im}[z]$ gives

$y = \sin \omega t/(25 - \omega^2)$. The general solution of the driven ODE $y'' + 25y = \sin \omega t$ is, for $\omega \neq 5$,

$$y = C_1 \cos 5t + C_2 \sin 5t + \frac{1}{25 - \omega^2} \sin \omega t$$

See Figs. 9(a), Graphs 1 and 2 for the orbits of the solutions where $y(0) = y'(0) = 0$, $\omega = 4$ and then $\omega = 1$. In Fig. 9(a), Graph 3, we have superimposed two orbits by writing a new system

$$x' = y$$
$$y' = -25x + \sin(zt)$$
$$z' = 0$$

and using initial data $x = 0$, $y = 0$, $z = 4, 1$. You see why we love to call this the "heart and eyes problem."

Problem 9, Graph 1.

Problem 9, Graph 2.

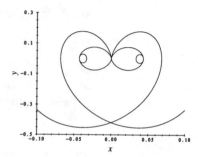

Problem 9, Graph 3.

11.   **(a).** If $z_d(t)$ is a solution of the ODE $z' + az = Ae^{i\omega t}$, then, since the coefficients in the operator $P(D) = D + a$ are real numbers, it follows from Theorem 3.6.3 that $y_d = \text{Re}[z_d]$ solves the same ODE but with the driving term replaced by $\text{Re}[Ae^{i\omega t}] = A \cos \omega t$.

(c). Writing

$$y_d = \text{Re}[z_d] = \frac{Aa}{a^2 + \omega^2} \cos \omega t + \frac{A\omega}{a^2 + \omega^2} \sin \omega t$$

and choosing an angle φ as indicated we have

$$y_d = \frac{A}{\sqrt{a^2 + \omega^2}} \left\{ \frac{a}{\sqrt{a^2 + \omega^2}} \cos \omega t + \frac{\omega}{\sqrt{a^2 + \omega^2}} \sin \omega t \right\}$$

$$= \frac{A}{\sqrt{a^2 + \omega^2}} \{\cos \varphi \cos \omega t + \sin \varphi \sin \omega t\}$$

$$= \frac{A}{\sqrt{a^2 + \omega^2}} \cos(\omega t - \varphi)$$

where we have used the formula $\cos(\alpha - \beta) = \cos \alpha \cos \beta + \sin \alpha \sin \beta$.

(e). We leave this to the group.

3.7 The General Theory of Linear ODEs

Comments

The two main results in this section, the General Solution Theorem and the Fundamental Theorem for Linear ODEs, are mostly of theoretical value, but form a nice way to pull things together as the chapter closes. If you intend to read about series methods in Chapter 11, then we suggest that you at least cover the highlights of this section. We made the Method of Variation of Parameters into a long group problem which outlines the development of the method. Although it's nice to know the statement of the Method, it's a lot of work to do examples where the driving term is not a polynomial-exponential.

So we concentrate on constant-coefficient operators $P(D)$ and polynomial-exponential driving terms. This covers most of the important applications in a first ODE course.

---

1. **(a).** Since $P(D)[y] = (tD^2 + D)[y] = ty'' + y' = (ty')'$, the null space of $tD^2 + D$ is found by integrating $(ty')' = 0$ to obtain $ty' = c_1$, so $y' = c_1 t^{-1}$. Integrating again, $y = c_1 \ln t + c_2$, $0 < t < \infty$, c_1 and c_2 arbitrary constants, which gives all the functions $y(t)$ in the null space.

   (c). In normal linear form we have that $y'' + y'/t = 2$, and the coefficient $1/t$ of y' is discontinuous on every interval containing the origin. There is no way to define the coefficient at $t = 0$ to make it continuous. So Theorem 3.7.1 does not apply, and there is no contradiction.

3. The functions y_1 and y_2 are shown to be solutions by direct substitution in the given ODE. The solution satisfying the given initial data is found by imposing the initial conditions on $y = c_1 y_1(t) + c_2 y_2(t)$ to determine c_1 and c_2. Since the normalized ODEs have discontinuous coefficients at $t = 0$, Theorem 3.7.1 applies only to intervals not containing 0.

   (a). The IVP is $t^2 y'' + ty' - y = 0$, $y(1) = 0$, $y'(1) = 0$. We have $y = c_1 t^{-1} + c_2 t$, with $y(1) = 0 = c_1 + c_2$ and $y'(1) = -1 = -c_1 + c_2$. So $c_1 = 1/2 = -c_2$ and $y = (t^{-1} - t)/2$, $t > 0$. See Fig. 3(a).

   (c). The IVP is $t^2 y'' + ty' - 4y = 0$, $y(0) = 0$, $y'(0) = 0$. In this case, $y = c_1 t^2 + c_2 t^{-2}$, but the initial data are given only at $t = 0$ where Theorem 3.7.1 is not applicable. If $c_2 = 0$, then the solution $y = c_1 t^2$ satisfies both initial conditions for all constants c_1. See Fig. 3(c) for the graphs of several solutions.

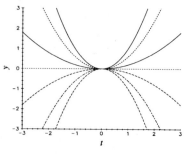

Problem 3(a).   Problem 3(c).

5.   Suppose $y = u(t)$ is a solution of $y'' + a(t)y' + b(t)y = 0$.

(a). Let $z(t)$ satisfy $uz'' + (2u' + au)z' = 0$. Then if $y = uz$ is substituted into $y'' + ay' + by = 0$, we have $(u''z + 2u'z' + uz'') + a(u'z + uz') + buz = uz'' + z'(2u' + au) + z(u'' + au' + bu) = 0$ since $u'' + au' + bu = 0$ and $uz'' + (2u' + au)z' = 0$. So $y = u(t)z(t)$ is a solution of $y'' + a(t)y' + b(t)y = 0$.

(c). We have that $W[u, uz] = u(uz)' - (uz)u' = u^2z' = u^2v = \exp[-\int^t a(r)\,dr] > 0$ where we have used $z' = v$ and v from part **(b)**, $v(t) = [u(t)]^{-2}\exp[-\int^t a(r)\,dr]$. Since the Wronskian is not 0 and $\{u, uz\}$ is a set of solutions of the normal ODE $y'' + ay' + by = 0$, the set $\{u, uz\}$ is a basic solution set.

7.   Let $\{u(t), v(t)\}$ be a basic solution set for the ODE, $y'' + a(t)y' + b(t)y = 0$. Let c and d be consecutive zeros of $u(t)$. We will show first that $v(t)$ has at least one zero between c and d. Let $W(t) = W[u, v](t)$ be the Wronskian of the basic solution pair $\{u, v\}$. Since $W(t) \neq 0$ for all t we may as well assume that $W(t) > 0$ [if $W(t) < 0$, then an analogous argument works]. So, $0 < W(c) = -v(c)u'(c)$, and $0 < W(d) = -v(d)u'(d)$, where we have used $u(c) = u(d) = 0$. Now from the Vanishing Data Theorem 3.7.2 we see that $u'(c) \neq 0$ and $u'(d) \neq 0$ since $u(t)$ and $v(t)$ are not identically 0. Since $0 < -v(c)u'(c)$ and $0 < -v(d)u'(d)$ and since c and d are consecutive zeros of u, it must be that $u'(c)$ and $u'(d)$ are of opposite sign. It follows that $v(c)$ and $v(d)$ are of opposite sign, so v must have a zero at some point between c and d. A similar argument shows that $u(t)$ must have at least one zero between any two consecutive zeros of $v(t)$. Therefore, between any two consecutive zeros of one solution there is precisely one zero of the other solution. That's why the two curves sketched in the margin *cannot* form a basic solution set for the ODE, $y'' + a(t)y' + b(t)y = 0$.

CHAPTER

4

Applications of Second-Order Differential Equations

There are two main applications in this chapter: the pendulum and electrical circuits. The pendulum models reappear numerous times in later chapters. In Section 4.1, a derivation is given for the model ODE of the simple pendulum starting from Newton's Laws of Motion. In Section 4.4 Kirchhoff's Voltage and Current Laws are used to derive ODEs that model current and charge in electrical circuits.

The material in Section 4.2 is a general treatment of beats and resonance for linear second-order ODEs with constant coefficients. We have written this section so that it can be covered independently of the applications sections. Engineering students may already have seen frequency response modeling in a systems or lumped parameters course, so the material of Section 4.3 will not come as a big surprise to them. Our approach builds on the properties of linear ODEs which were developed in Chapter 3.

Periodic forced oscillations have been touched upon at various places in this text other than in Sections 4.2 and 4.3. Problem 12 in Section 2.2 outlines a proof of the existence of a unique periodic forced oscillation for a first-order, periodically driven and linear ODE, and (for good measure) we give a simpler version in Group Problem 5 in Section 4.2. Group Problem 9 in Section 4.3 does the same for second-order linear ODEs, as does the group problem in Section 7.9 for linear systems.

4.1 Newton's Laws: The Pendulum

Comments

The proper language for Newtonian mechanics is the language of vectors. We have avoided vectors up to this point by considering motion only in one dimension—for example, a falling body, a vertically oscillating spring. But at some point or other vectors should be introduced so that Newton's three laws of motion can be fully understood. The approach to vectors in this section is traditional, and nonaxiomatic. Vectors are thought of as arrows, but denoted by boldface type. The dot product is introduced and some of its properties are listed (Problem 1). Unit coordinate vectors $\mathbf{i}$, $\mathbf{j}$, $\mathbf{k}$, $\hat{\mathbf{x}}$, $\hat{\mathbf{y}}$, $\hat{\mathbf{z}}$, $\hat{\mathbf{r}}$, $\hat{\boldsymbol{\theta}}$ are used. The main application of the vector approach is the derivation of the equation of motion of the simple pendulum. We have opted to leave the derivation of various properties of the solutions and orbits of the simple undamped pendulum (both nonlinear and linearized) to the problem set (Problems 5–8).

1. **(a).** Let $\mathbf{u} = (u_1, u_2, u_3)$. The value of $\|\mathbf{u}\|^2$ is the square of the distance from the origin $(0, 0, 0)$ to the point (u_1, u_2, u_3). So, $\|\mathbf{u}\|^2 = \left(\sqrt{(u_1 - 0)^2 + (u_2 - 0)^2 + (u_3 - 0)^2}\right)^2 = u_1^2 + u_2^2 + u_3^2$ for any vector $\mathbf{u}$.

    (c). Let $\mathbf{u} = (u_1, u_2, u_3)$ and $\mathbf{v} = (v_1, v_2, v_3)$. Then $\mathbf{u} \cdot \mathbf{v} = u_1 v_1 + u_2 v_2 + u_3 v_3 = v_1 u_1 + v_2 u_2 + v_3 u_3 = \mathbf{v} \cdot \mathbf{u}$.

    (e). Let $\mathbf{u} = (u_1, u_2, u_3)$. Then $\mathbf{u} \cdot \mathbf{u} = u_1^2 + u_2^2 + u_3^2 \geq 0$ for all $\mathbf{u}$. Since a sum of squares of real numbers is 0 if and only if each number is 0, then $\mathbf{u} \cdot \mathbf{u} = u_1^2 + u_2^2 + u_3^2 = 0$ if and only if $\mathbf{u} = \mathbf{0}$.

3. Applying Newton's Second Law (in the vertical direction) to the piston, we have $mx'' =$ force of gravity $+$ force due to gas pressure. The magnitude of the force due to gas pressure is $P \cdot A$ where A is the area of the piston. But $P = nRT/V = nRT/(x \cdot A)$, so the force due to gas pressure is $P \cdot A = nRT/x$. Since the force of gravity is directed downward while the force due to gas pressure is directed upward, $mx'' = -mg + nRT/x$.

5. **(a).** See Fig. 5(a) for some of the orbits of the ODE $mL\theta'' + cL\theta' + mg \sin \theta = 0$ with $m = 1$, $c = 0$, and $g/l = 10$, including the over-the-top orbits (wavy lines at top and bottom) and the back-and-forth orbits seen here as ovals. Can you plot the separatrices (we didn't here, but did in text Figure 4.1.1)?

    (c). See Fig. 5(c), Graph 1 for orbits of the linearized, undamped and undriven pendulum ODE, $\theta'' + 10\theta = 0$. The orbits are the ellipses $(\theta')^2 + 10\theta^2 =$ constant. Close to the equilibrium point $\theta = 0$, $\theta' = 0$, these ellipses resemble the oval orbits of the nonlinear ODE, $\theta'' + 10 \sin \theta = 0$. But at a distance from the point $\theta = 0$, $\theta' = 0$ the orbits of the linearized ODE don't look like the orbits of the nonlinear ODE at all. The linear ODE has only the single equilibrium point at the origin; the nonlinear ODE has infinitely many equilibrium points. All nonconstant orbits of the linear ODE are closed curves that surround the origin, and that is certainly not true for the nonlinear ODE. So we see that the linearized ODE, $\theta'' + 10\theta = 0$, may be a pretty good approximate model of the simple undamped pendulum dynamics near the origin, but is a total failure elsewhere, failing completely to model the other equilibrium points, over-the-top motions, and so on.

See Fig. 5(c), Graph 2 for orbits of the linearized, damped and undriven pendulum ODE, $\theta'' + \theta' + 10\theta = 0$. The inward spiraling orbits resemble the central inward spiraling orbits of the chapter cover figure, but at a distance from the equilibrium point $\theta = 0$, $\theta' = 0$, this linearized ODE is a terrible approximation to the nonlinear ODE, $\theta'' + \theta' + 10\sin\theta = 0$ for many of the same reasons mentioned in the preceding paragraphs.

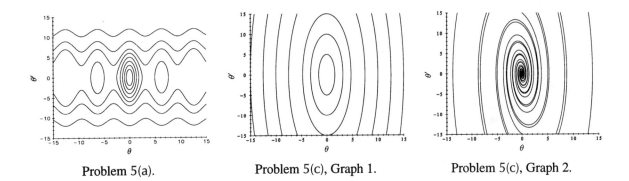

Problem 5(a).   Problem 5(c), Graph 1.   Problem 5(c), Graph 2.

7. **(a).** Multiplying the ODE $mL\theta'' + mg\sin\theta = 0$ by θ', we have $mL\theta'\theta'' + mg\theta'\sin\theta = 0$. Now $mL\theta'\theta'' = [mL(\theta')^2/2]'$ and $mg\theta'\sin\theta = [-mg\cos\theta]'$, so the ODE becomes $mL\theta'\theta'' + mg\theta'\sin\theta = [mL(\theta')^2/2 - mg\cos\theta]' = 0$, and we have that $mL(\theta')^2/2 - mg\cos\theta = c$, where c is a constant. See the plots in Fig. 5(a) for some of these orbits.

9. Group project. Here is a sketch of the solution of the bug problem (Group Problem 9). Orient the table so that the center is at the origin and the x- and y-axes lie along the diagonals of the table. As the bugs move, these positions form the corners of a rotating and shrinking square. Since the problem is symmetric with respect to the paths of the four bugs, we will only find the path of one of the bugs, bug A say. Let $\mathbf{R}(t) = r(t)\hat{\mathbf{r}}$ be the position vector of bug A, and let $\hat{\boldsymbol{\theta}}$ be the unit vector orthogonal to $\hat{\mathbf{r}}$ and pointing toward bug B, the bug adjacent to A in the counterclockwise sense. Because of the symmetries of the problem, the velocity vector $\mathbf{R}'(t)$ always makes an angle of $135°$ with $\hat{\mathbf{r}}$ and an angle of $45°$ with $\hat{\boldsymbol{\theta}}$. But this implies that the component of $\mathbf{R}'(t)$ along $\hat{\mathbf{r}}$ is equal in magnitude but opposite in sign to the component along $\hat{\boldsymbol{\theta}}$. Since $\mathbf{R}' = r'\hat{\mathbf{r}} + r\hat{\mathbf{r}}' = r'\hat{\mathbf{r}} + r\theta'\hat{\boldsymbol{\theta}}$, the components referred to are r' and $r\theta'$, respectively. We must have $r' = -r\theta'$, that is, $dr/d\theta = -r$, and $r = Ce^{-\theta}$. Since $r = a/\sqrt{2}$ at $\theta = 0$, it follows that $C = a/\sqrt{2}$, and so we have $r = ae^{-\theta}/\sqrt{2}$. Considering the bugs as dimensionless points, we see that the path of each bug involves an infinitely increasing θ; $r \to 0$ as $\theta \to +\infty$.

And now for the kicker: although the angle θ goes to ∞, it takes only a finite time for the four bugs to meet at the center of the table! This follows from the fact that the speed, $\|\mathbf{R}'(t)\| = ((r')^2 + (r\theta')^2)^{1/2}$ is a positive constant, say b. Since $r' = -r\theta'$ (as above), we see that $b^2 = (r')^2 + (r\theta')^2 = 2(r')^2$. So, $r' = -b/\sqrt{2}$ (remember that $r' < 0$). Using the initial data $r(0) = a/\sqrt{2}$, we have that $r(t) = (-bt + a)/\sqrt{2}$. When $t = a/b$, $r(t) = 0$

and the bugs meet, probably pretty dizzy after all that whirling around the origin. Since $\theta' = -r'/r = -(\ln r)' = [\ln(1/r)]'$, we have that

$$\theta = \ln \left(\frac{1}{1 - (b/a)t} \right)$$

and we see that $\theta \to +\infty$ as $t \to (b/a)^-$.

4.2 Beats and Resonance

Comments

In this section we see how second-order constant coefficient linear ODEs respond to a sinusoidal driving force. ODEs of this kind often appear in the sciences and engineering. The notions of beats, resonance, free and forced oscillations (both in the undamped and damped cases) all arise in this setting. Take a pair of tuning forks tuned to slightly different frequencies and listen to the beats.

1. In each part, the equation has the form $y'' + \omega_0^2 y = A \cos \omega t$.

(a). For the ODE $y'' + 9y = 5\cos 2t$, $A = 5$, $\omega_0 = 3$, and $\omega = 2$, so the solution exhibits beats. By (11) of Case 2, the general solution is $y(t) = C_1 \cos 3t + C_2 \sin 3t + \cos 2t$. Since $y(0) = y'(0) = 0$, $C_1 = -1$ and $C_2 = 0$, so the solution is $y(t) = \cos 2t - \cos 3t$.

(c). For the ODE $y'' + y = \cos t$, $A = 1$ and $\omega_0 = \omega = 1$, so resonance occurs. By Case 1 in the text, $y = C_1 \cos t + C_2 \sin t + (1/2)t \sin t$. Since $y(0) = 1$ and $y'(0) = 0$, $C_1 = 1$ and $C_2 = 0$, so the solution is $y = \cos t + (1/2)t \sin t$.

3. Solving the IVP $y'' + \omega_0^2 y = A \sin \omega t$, $y(0) = 0$, $y'(0) = -A/(\omega_0 + \omega)$ as in Case 2 in the text, we obtain the general solution

$$y(t) = C_1 \cos \omega_0 t + C_2 \sin \omega_0 t + \frac{A}{\omega_0^2 - \omega^2} \sin \omega t$$

Since $y(0) = 0$, we have that $C_1 = 0$. Using $y'(0) = -A/(\omega_0 + \omega)$, we have $C_2 \omega_0 + A\omega/(\omega_0^2 - \omega^2) = -A/(\omega_0 + \omega)$, so $C_2 = -A/(\omega_0^2 - \omega^2)$. So

$$y(t) = A/(\omega_0^2 - \omega^2)[\sin \omega t - \sin \omega_0 t]$$

Using the hint,

$$y(t) = [2A/(\omega_0^2 - \omega^2)] \sin ((\omega - \omega_0)t/2) \cos ((\omega + \omega_0)t/2)$$

which exhibits the phenomenon of beats [see the explanation following (12)].

5. Here is the basic approach to the solution of the group problem. See Figs. 5, Graphs 1 and 2 for the solution curves of the two IVPs, $y'' + 4y = \cos \omega t$, $y(0) = y'(0) = 0$, $\omega = 3$, π. Figure 5, Graph 1 corresponds to $\omega = 3$, Fig. 5, Graph 2 to $\omega = \pi$. Only the upper part of the solution curve is plotted for $\omega = \pi$ so that the non-periodicity of the solution is clearly visible. The natural frequency for both solutions is $\omega_0 = 2$. The first solution curve is periodic of period 2π, but you can't easily detect the forced oscillation of period $2\pi/\omega = 2\pi/3$. The second curve is not periodic at all. In general, the solutions are given by formula (11) with $\omega_0 = 2$, $\omega = 3$, π: $y(t) = C_1 \cos 2t + C_2 \sin 2t + A/(4 - \omega^2) \cos \omega t$. If $\omega = 3$, then all solutions have period 2π as noted. If $\omega = \pi$, then the ratio of ω_0 and ω is irrational, and there are no periodic solutions unless $C_1 = C_2 = 0$. What do these results imply about the problem of detecting a periodic forced oscillation by examining the graph of one solution of $y'' + \omega_0 y = A \cos \omega t$?

Problem 5, Graph 1.

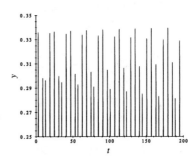

Problem 5, Graph 2.

7. Group project.

4.3 Frequency Response Modeling

Comments

Frequency response modeling is a meat-and-potatoes item on the engineering menu. Mathematicians are less accustomed to seeing the damped resonance phenomena couched in the engineering language of gain, phase shift, and Bodé plots used here, but once we get used to the new terms they seem natural and down-to-earth. The only real problem that we see is not the new terminology, but the need for care in the algebraic manipulation. We also touch on one of the most important engineering problems, the problem of parameter identification: How can you determine the coefficients of an ODE from a knowledge of the input and output?

1. The text finds the unique periodic solution (11) of $y'' + 2cy' + k^2 y = F_0 \sin \omega t$. Proceeding in an analogous manner, replacing $\sin \omega t$ with $\cos \omega t$, but this time extracting the real instead of the imaginary part of $F_0 M(\omega) \exp[i(\omega t + \varphi(\omega))]$, we obtain

$$y_p = F_0 M(\omega) \cos(\omega t + \varphi(\omega))$$

where $M(\omega) = [(k^2 - \omega^2)^2 + 4c^2\omega^2]^{-1/2}$ and $\varphi(\omega) = \cot^{-1}\left((\omega^2 - k^2)/2c\omega\right)$ with $-\pi \le \varphi(\omega) \le 0$. See the discussion in the text following the second equation in (10) for the reason why $-\pi \le \varphi(\omega) \le 0$.

3. Here, using the notation from ODE (1), $y'' + 2cy' + k^2y = F_0 \sin \omega t$, we have $2c = 1$ and $k^2 = 2$ for the ODE, $y'' + y' + 2y = F_0 \sin \omega t$. So, from (10) we have that

$$M(\omega) = \frac{1}{\sqrt{(2-\omega^2)^2 + \omega^2}} = \frac{1}{\sqrt{4 - 3\omega^2 + \omega^4}}, \quad \varphi(\omega) = \cot^{-1}\left(\frac{\omega^2 - 2}{\omega}\right)$$

(a). $M(\omega)$ and $\varphi(\omega)$ can be found graphically by solving the ODE for various values of ω. $M(\omega)$ and $\varphi(\omega)/\omega$ can then be read off the plots. Plots should be similar to the plot given in Fig. 4.3.4.

5. (a). The Hooke's Law spring constant k satisfies the relationship $k \times$ deflection $= mg$. Using $g = 9.8$ m/s^2, $m = 1$ kg, deflection $= 0.0127$ m, we find $k \approx 772N/m$. Solutions of $y'' + 2cy' + ky = 0$, where $c^2 < k$ so that the roots of the characteristic polynomial have non-zero imaginary part, have the form $y = Ae^{-ct}\cos\sqrt{k - c^2}t + Be^{-ct}\sin\sqrt{k - c^2}t$. Amplitudes decay by a factor of $0.01016/0.00254 = 4$ in 20 cycles here, so we need $e^{-c(20T)} = 1/4$, that is, $T = (1/20c)\ln 4$. Now $T = 2\pi/\sqrt{k - c^2}$, so we obtain $c = (1/40\pi)(\ln 4)\sqrt{k - c^2}$. Solving for c with $k = 772N/m$, we obtain $2c = 0.613/s$.

7. Group project.

9. Group project.

4.4 Electrical Circuits

Comments

Electrical circuits provide a very nice application of second-order constant coefficient linear ODEs, an application of immense importance in today's technological world of communication channels. We start from scratch here and talk about charge carriers, voltage, currents, the whole bit. The going is heavy at times, but no one says that electricity is easy to understand. When all of the characters in the electrical circuit story have been introduced, we turn to the simple RLC circuit and solve the corresponding model ODE for the current in the circuit—no new solution techniques here, just the new language of circuits. The solution calculations are neither easier nor harder than before.

---

1. (a). From IVP (5), $LI'' + RI' + I/C = E'(t)$, we have, assuming that $R = 20$, $L = 10$, $C = 0.05$, $E(t) = 12$, and $10I'' + 20I' + I/0.05 = E' = 0$, or $I'' + 2I' + 2I = 0$. In addition, inserting $q_0 = 0.6$, $I_0 = 1$ into the boundary data formulas given in (5), we have $I(0) = 1$, $I'(0) = -2$. Since the roots of the characteristic polynomial are $-1 \pm i$, the general solution is $I(t) = k_1e^{-t}\cos t + k_2e^{-t}\sin t$. Applying the initial conditions, $k_1 = 1$ and $k_2 = -1$. So, $I(t) = e^{-t}[\cos t - \sin t]$, which is oscillatory with circular frequency 1 and decaying amplitude.

(c). Overdamping occurs when the characteristic polynomial $Lr^2 + Rr + 1/C$ has distinct real roots, that is, if $R^2 > 4L/C$. In this case, that means when $R^2 > 800$, that is, when $R > 20\sqrt{2}$ ohms.

3. Let q_1 denote the charge on the capacitor C_1, q_2 the charge on the capacitor C_2. Then by Kirchhoff's Voltage Law for the left and the right loops, respectively, $q_1/C_1 + q_2/C_2 = 0$ and $Rq_2' + q_2/C_2 = 0$, where $q_1(0) = C_1 E_0$.

    The IVP for $q_2(t)$ is $q_2' = -q_2/RC_2$, $q_2(0) = -C_2 E_0$, and so $q_2(t) = -C_2 E_0 e^{-t/RC_2}$. So $q_1 = -(C_1/C_2)q_2 = C_1 E_0 e^{-t/RC_2}$ and $I_1(t) = q_1'(t) = -C_1 E_0 e^{-t/RC_2}/C_2 R = (-1/6) \times 10^{-6} E_0 e^{-t/6}$ since $C_1 = 10^{-6}$, $C_2 = 2 \times 10^{-6}$, $R = 3 \times 10^6$.

5. According to IVP (5), we have $2I'' + 7 \times 10^4 I' + 4I = E' = (60)' = 0$, $I(0) = 0$, $I'(0) = 60/2 = 30$. The solution is $I(t) = k_1 e^{r_1 t} + k_2 e^{r_2 t}$, where r_1 and r_2 are the roots of $r^2 + 3.5 \times 10^4 r + 2 = 0$ and $0 = k_1 + k_2$, $30 = r_1 k_1 + r_2 k_2$. Carrying out the calculations, we have $k_1 = 30/(r_1 - r_2) = -k_2$, and $r_1 = -34999.99994$, while $r_2 = -0.00006$. So, $k_1 \approx -0.000857 = -k_2$. The charge $q(t) = q_0 + \int_0^t I(s)\,ds = (k_1)/(r_1)[e^{r_1 t} - 1] - (k_1)/(r_2)[e^{r_2 t} - 1]$ coulombs. The dominant term is $-(k_1)/(r_2)[e^{r_2 t} - 1] \approx -14.29[e^{-0.00006t} - 1]$. If $t = 0.1$ sec, $q(0.1) \approx 0.000086$ coulombs.

7. With $L = 1$, $C = 1/25$ we know that the tuned circuit will home in on the frequency $\omega = 5$ input signal since $\omega^2 = 25 = 1/LC$ [see (16)]. The question is what will the various values of the resistance do to the amplitude of the response? See Fig. 7, Graph 1 for the graphs of the current vs. time in the three cases $R = 1.0$ (solid), $R = 0.1$ (short dashes), and $R = 0.01$ (long dashes). Apparently the lower the resistance, the larger the amplitude of the response; in the language of Section 4.3, the greater the gain. We have tuned the circuit to the input frequency ω by choosing C so that $1/LC = \omega^2$, and the resulting steady-state current is $I_p = (E_0/R)\cos(\omega t + \varphi)$ where $\omega E_0 = 4$ so $E_0 = 4/\omega = 4/5$. Since the amplitude of the input for the current ODE is ωE_0 (*not* E_0, which is the amplitude of the charge ODE), the gain is $(E_0/R)/E_0\omega = 1/R\omega$, i.e., $1/(5R)$ since $\omega = 5$. The amplitudes of the response currents seen in Fig. 7, Graph 1 approach $E_0/R = 4/5$ when $R = 1$ (short dashes), $(4/5)/(0.1) = 8$ when $R = 0.1$ (long dashes), and $(4/5)/(0.01) = 80$ when $R = 0.01$ (solid). The graphs are not of I_p, but include some transients as well. As t increases the graphs approach the graph of I_p. This is particularly evident if $R = 1.0$ and 0.1, where over the interval $90 \le t \le 100$, we are essentially looking at $I_p(t)$ with respective amplitudes $4/5$ ($R = 1.0$), 8 ($R = 0.1$). But, the graph corresponding to $R = 0.01$ has a long way to go before its amplitude approaches 80. Fig. 7, Graph 2 shows the current if $R = 0.01$; the ODE is solved over $0 \le t \le 1000$, but plotted only for $975 \le t \le 1000$, by which point the amplitude has approached 80.

Problem 7, Graph 1.

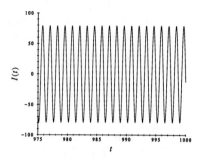

Problem 7, Graph 2.

9.   Group project.

Systems of
Differential Equations

Systems of ODEs have been used in earlier chapters, but informally. In Chapters 5, 7, 8, and 9 we describe the theory and practice of systems. Since solution formulas for systems of ODEs (particularly when they are not linear) are even rarer than they are for first- or second-order ODEs, we start out with computer simulations. That way, you can see what component curves, time-state curves, orbits, and so on look like. Now for some general observations about the contents of Chapter 5. Many models are presented in this chapter: coupled springs and masses, chemical reactions, models of predator-prey dynamics, competing or cooperating species, and models of the spread of a disease. The aim of this profusion of models is to show the naturalness of systems of ODEs as models of natural processes. Our experience is that people take to systems right away if there are models around to give some physical reality to the mathematical constructs. The chapter cover figure shows a time-state orbit of a model system for an autocatalytic reaction (see Example 5.1.5).

As we have done in Chapters 2 and 3, we introduce a version of the Fundamental Theorem (Theorem 5.2.1) for the existence, uniqueness, extension, and continuity/sensitivity properties of the solutions of initial value problems for systems of ODEs. Since this is its third appearance, readers will know what to expect, and have an idea of why we keep talking about these basic ideas. The sensitivity of solutions to changes in data again comes up in the text and in the problems. Our goal is to hit this idea so often that everyone will begin to understand its importance in the applications of ODEs. This is an area where computer simulations are a critical factor in conveying the effects of parameter changes.

Population models (like the Lotka-Volterra models of Section 5.4) are controversial because they are often viewed as reductionist; that is, they vastly oversimplify the reality of population dynamics. And that is certainly true in many cases. However, these models do provide a way to think about interacting species, and they are the first step in most attempts to gain an understanding of how populations interact. The possum disease model of Section 5.5 is due to Professor Graeme Wake and his colleagues in New Zealand. We have included it here because it is still in the development stage and models an important problem for the New Zealand economy. We think that every ODE course should have at least one open question, model, or theorem—something where the final "answer" isn't yet known. The possum model is one of these.

5.1 First-Order Systems

Comments

Systems of first-order ODEs have appeared several times in models in earlier chapters. In this section we show that a system of springs and masses may by modeled by a system of four linear first-order ODEs; the state variables in the model are the positions and the velocities of the masses suspended by the springs. The basic modeling elements for Hooke's Law springs appear in Section 3.1 and are used in constructing the model. We will come back to this model in Chapter 7, where we introduce the terminology of normal modes and natural frequencies to describe the periodic oscillations. We also introduce a new kind of model in the autocatalator system—a model of chemical kinetics. The basic principle of Chemical Mass Action is given and discussed; from the principle we can write out the rate equations for each of the species involved in the chemical reaction. The rate equations for autocatalytic reactions are nonlinear, and their solutions behave quite strangely.

The examples in this section are discussed entirely in the light of their computer simulations. The aim of the examples is to show the naturalness of systems in modeling phenomena and to give the user of this book experience with computer simulation of systems. We also introduce vector notation for modeling systems, but we do not use the boldface notation of Section 4.1. Component curves, orbits, and time-state curves are defined and graphed. Linear systems are defined, but their full treatment is delayed to Chapter 7.

Experiments

Both of the processes described in this section can be illustrated with experiments. Physicists can give you springs and masses that can slide back and forth horizontally along a smooth surface. Then you can experiment with initial displacements from equilibrium that will give in-phase, out-of-phase oscillations, and non-periodic motions.

You can also illustrate the autocatalytic chemical process for yourself if you can line up some chemical equipment. If you want to do this experiment, take a look at *Chemical Demonstrations* by Lee R. Summerlin and James L. Ealy; Americal Chemical Society, Washington, 1985. Summerlin and Ealy give you all the information you need to do three autocatalytic experiments. In each experiment the autocatalytic oscillations are seen as color changes in a chemical solution. The actual reactions taking place are quite complicated, so the model given in this section can be viewed as just a rough "caricature" of what is going on in an actual reaction.

---

1. **(a).** The characteristic polynomial $r^2 - 4$ for $y'' - 4y = 0$ has roots ± 2, so the general solution is $y = C_1 e^{-2t} + C_2 e^{2t}$. If $x_1 = y$ and $x_2 = y'$, the equivalent system is $x_1' = x_2$, $x_2' = 4x_1$ with general solution $x_1 = y = C_1 e^{-2t} + C_2 e^{2t}$, $x_2 = y' = -2C_1 e^{-2t} + 2C_2 e^{2t}$.

(c). The characteristic polynomial $r^2 + 5r + 4$ for $y'' + 5y' + 4y = 0$ has roots $-4, -1$, so the general solution is $y = C_1 e^{-4t} + C_2 e^{-t}$. The equivalent system is $x_1' = x_2$, $x_2' = -4x_1 - 5x_2$ with general solution $x_1 = y = C_1 e^{-4t} + C_2 e^{-t}$, $x_2 = y' = -4C_1 e^{-4t} - C_2 e^{-t}$.

(e). Let $x_1 = y$, so $x_2 = x_1' = y'$ and $y'' = x_2' = -16x_1 = -16y$. The scalar ODE that is equivalent to the system, $x_1' = x_2$, $x_2' = -16x_1$, is $y'' + 16y = 0$. The characteristic polynomial is $r^2 + 16$, which has roots $\pm 4i$. The general solution is $y = C_1 \cos 4t + C_2 \sin 4t$. The general solution of the system is $x_1 = y = C_1 \cos 4t + C_2 \sin 4t$, $x_2 = y' = -4C_1 \sin 4t + 4C_2 \cos 4t$.

3. These linear cascade problems are all solved by first solving the "source ODEs" and inserting the solutions in the ODEs at the next cascade level, solving those ODEs and repeating the process until the "final products" have been found. In every case we use the integrating factor technique of Sections 1.3 and 1.4.

(a). The solution of the first IVP in the system $x_1' = -3x_1$, $x_2' = x_1$, $x_3' = 2x_1$, $x_1(0) = 0$, $x_2(0) = x_3(0) = 0$, is $x_1 = 10e^{-3t}$. Inserting this solution for x_1 into the second and third equations, $x_2' = 10e^{-3t}$ and $x_3' = 20e^{-3t}$. Integrating these and applying the initial conditions, we obtain the solution $x_1 = 10e^{-3t}$, $x_2 = 10(1 - e^{-3t})/3$, $x_3 = 20(1 - e^{-3t})/3$. The limiting values as $t \to +\infty$ are $x_1 = 0$, $x_2 = 10/3$ and $x_3 = 20/3$. The source variable is x_1; x_2 and x_3 are final products.

(c). The solutions of the first and second IVPs in the system $x_1' = -2x_1$, $x_2' = -3x_2$, $x_3' = 2x_1 + 3x_2$, $x_1(0) = 1$, $x_2(0) = 2$, $x_3(0) = 0$ are $x_1 = e^{-2t}$, $x_2 = 2e^{-3t}$. Inserting these into the third equation, $x_3' = 2e^{-2t} + 6e^{-3t}$. Integrating and applying the initial conditions, we obtain the solution $x_1 = e^{-2t}$, $x_2 = 2e^{-3t}$, $x_3 = 3 - e^{-2t} - 2e^{-3t}$. The limiting values as $t \to +\infty$ are $x_1 = x_2 = 0$, $x_3 = 3$; x_1 and x_2 are source variables, and x_3 is the final product variable.

5. **(a).** For the system $x_1' = x_2$, $x_2' = -3x_2$, $f_1 = x_2$ and $f_2 = -3x_2$. Separating variables in the ODE, $dx_2/dx_1 = f_2/f_1 = -3x_2/x_1$, we have $dx_2/x_2 = -3dx_1/x_1$, implying that $\ln|x_2| = -3\ln|x_1| + K$. So $x_2 = Cx_1^{-3}$ is a family of solutions of $dx_2/dx_1 = -3x_2/x_1$ and gives a family of orbits of the original system. Note that $x_1 = 0$ is excluded because of the division by x_1 in the ODE, although $x_1(t) = 0$, for all t, is an orbit of the original system. This shows that some orbits may be lost by this solution process. See Fig. 5(a).

(c). For the system $x_1' = x_2$, $x_2' = -e^{-x_1}$, $f_1 = x_2$ and $f_2 = -e^{-x_1}$. We have that $dx_2/dx_1 = -e^{-x_1}/x_2$. Separating variables and solving, we obtain $x_2^2 = 2e^{-x_1} + C$; note that for each negative C, x_1 must be restricted to an interval for which $2e^{-x_1} + C \geq 0$, that is, $x_1 \leq \ln(-2/C)$. Note also that $x_2(t)$ is always decreasing because x_2' is negative. Since $x_2' = -e^{-x_1}$, the magnitude of x_2' is huge if x_1 is negative. See Fig. 5(c).

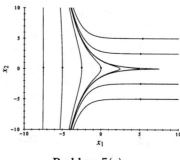

Problem 5(a).   Problem 5(c).

7. Use the discussion above text system (17) and also the Chemical Law of Mass Action as guides. Lowercase letters w, x, y, z denote concentrations.

(a). The speed of the first step of the reaction schematic, $X + Y \overset{k_1}{\to} Z \overset{k_2}{\to} W$, is $k_1 xy$, while the speed of the second step is $k_2 z$. So the ODEs are $x' = -k_1 xy$, $y' = -k_1 xy$, $z' = k_1 xy - k_2 z$, $w' = k_2 z$.

(c). The speed of the reaction with the schematic, $X + 2Y \overset{k}{\to} 6Y + W$, is kxy^2. Since 2 units of Y are consumed to generate 6 units of Y, the net gain is 4 units of Y. So the ODEs are $x' = -kxy^2$, $y' = 4kxy^2$, $w' = kxy^2$.

9. Group project.

5.2 Properties of Systems

Comments

Here we are with another Fundamental Theorem (Theorem 5.2.1). This time the theorem is so inclusive it contains the Fundamental Theorem of Section 2.3. Since we work mostly with systems of first-order ODEs in Chapters 5, 7, 8, and 9, the theorem is formulated for these systems. There is quite a bit of basic material for the study of systems of ODEs in this section besides the Fundamental Theorem: autonomous systems, equilibrium points and cycles, maximal extension of solutions of autonomous systems, planar autonomous systems and their direction fields and nullclines, sensitivity to parameter changes.

One very important point may get buried in the mass of detail: parameters in the rate functions can be treated as additional state variables whose rates of change are all zero. The text above Example 5.2.7 shows how this is done. Applying the Fundamental Theorem to the extended system, we conclude that solutions are continuous in all the parameters as long as the hypotheses of the theorem are satisfied by the rate functions.

1. The hypotheses of the Fundamental Theorem are satisfied in each part for all points in state space and for all t since the rate functions in each case are polynomials in the state variables, and all polynomials are continuous and continuously differentiable. So, each

of the initial value problems has a unique maximally extended solution defined on some time interval containing $t_0 = 0$. The solutions are defined for all real t unless specifically noted otherwise. The origin in state space is the only equilibrium point since the right-hand sides vanish only at that point. The x- and y-nullclines are the dashed curves. They are not always visible because sometimes orbits lie on top of them. Some of the nine orbits coincide with one another, and so fewer than nine orbits may be visible in the pictures. The 9 initial points to use are $(-1, -1)$, $(-1, 0)$, $(-1, 2)$, $(0, -1)$, $(0, 0)$, $(0, 2)$, $(2, -1)$, $(2, 0)$, and $(2, 2)$.

(a). The system is $x' = y$, $y' = -x - 2y$. The corresponding scalar IVP, $x'' + 2x' + x = 0$, $x(0) = a$, $x'(0) = y(0) = b$, has characteristic polynomial $r^2 + 2r + 1$ with -1 as a double root. The general solution has the form $x = C_1 e^{-t} + C_2 t e^{-t}$. Applying the initial conditions, we obtain $x = e^{-t}(a + (a + b)t)$. Then $y = x' = e^{-t}(b - (a + b)t)$. See Fig. 1(a), where the x-nullcline is $y = 0$, and the y-nullcline is $y = -x/2$; $(0, 0)$ is the equilibrium point.

(c). The system is $x' = y$, $y' = -9x$. The equivalent scalar IVP $x'' + 9x = 0$, $x(0) = a$, $x'(0) = y(0) = b$ has characteristic polynomial $r^2 + 9$ with roots $\pm 3i$. The general solution has the form $x = C_1 \cos 3t + C_2 \sin 3t$. Applying the initial conditions, we obtain $x = a \cos 3t + (b \sin 3t)/3$, and so $y = x' = -3a \sin 3t + b \cos 3t$. All orbits aside from the equilibrium point $(0, 0)$ are cycles. See Fig. 1(c), where the x- and y-nullclines are, respectively, $y = 0$ and $x = 0$.

(e). The ODEs $x' = -x^3$, $y' = -y$ are uncoupled and may be solved directly. Separating variables, we have $x^{-3}dx = -dt$, $x(0) = a$, which leads to $a^{-2} - x^{-2} = -2t$, or $x = a(1 + 2a^2 t)^{-1/2}$, where $t > -1/2a^2$ (or if $a = 0$, $|t| < \infty$). In addition, the solution of the IVP, $y' = -y$, $y(0) = b$, is $y = be^{-t}$ for the same t-interval. See Fig. 1(e), where the x- and y-nullclines are, respectively, $y = 0$ and $x = 0$; $(0, 0)$ is an equilibrium point.

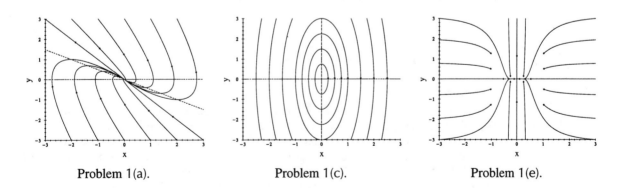

Problem 1(a).   Problem 1(c).   Problem 1(e).

3. The ODEs in parts **(a)**–**(c)** are uncoupled and may be solved separately. Notice that translating solutions by T just amounts to changing the values of the integration constants.

(a). Since the ODEs, $x' = 3x$, $y' = -y$, are uncoupled, the equations are easily solved, $x(t) = C_1 e^{3t}$, $y(t) = C_2 e^{-t}$, $-\infty < t < \infty$. Since $x(t - T) = C_1 e^{3(t-T)}$, we have that

$dx(t - T)/dt = 3C_1 e^{3(t-T)} = 3x(t - T)$, and so $x(t - T)$ is a solution of $x' = 3x$ if $x(t)$ is a solution. Similarly, $dy(t - T)/dt = -C_2 e^{-(t-T)} = -y(t - T)$, so $y(t - T)$ is a solution. The solution $x = x(t - T)$, $y = y(t - T)$ is defined for all t.

(c). Separating variables and solving the first ODE, $x' = -x^3$, for x, we have that $x = C_1(1 + 2C_1^2 t)^{-1/2}$, $t > -1/2C_1^2$ while the second ODE, $y' = 1$, has the solutions $y(t) = t + C_2$, $t > -1/2C_1^2$. Since $dx(t - T)/dt = -C_1^3(1 + 2C_1^2(t - T))^{-3/2} = -x^3(t - T)$, $x(t - T) = C_1(1 + 2C_1^2(t - T))^{-1/2}$, $t > -1/2C_1^2 + T$ is a solution. Similarly, $dy(t - T)/dt = 1$, so $x(t - T)$, $y(t - T)$, $t > -1/2C_1^2 + T$ is a solution of the system.

5.   **(a).** Suppose that P is an equilibrium point of $x' = f(x)$, and that P is inside a region S in which the conditions of the Fundamental Theorem hold. Then $x = P$ is a solution. Suppose $x = x^1(t)$, $a < t < b$, is another maximally extended solution of the system and that $x^1(t) \to P$ as $t \to T$. By the maximal extension property for the orbits of an autonomous system, and the fact that $x^1(t)$ is a continuous vector function of t, we have that either $a < T < b$ and $x^1(T) = P$, or else $T = a = -\infty$, or else $T = b = +\infty$, or else $T = a$ (or b) and $x^1(t)$ tends to the boundary of S as $t \to a^+$ (or b^-). The first alternative is forbidden by the uniqueness property since otherwise there would be two different maximally extended orbits through P, $x = P$ for all t and $x = x^1(t)$. The last alternative would imply that P lies on the edge of S, but P is assumed to lie inside S. So one of the two middle alternatives must hold, $T = \infty$ or $T = -\infty$, and a solution reaches a constant equilibrium state only as $t \to \infty$, or $t \to -\infty$.

7.   The nonconstant vector functions $\begin{bmatrix} \sin t \\ t \end{bmatrix}$ and $\begin{bmatrix} te^t \\ t^3 \end{bmatrix}$ are both the zero vector at $t = 0$.
So there are *no* continuously differentiable functions $f(x, y, t)$, $g(x, y, t)$ such that both vector functions are solutions of $x' = f$, $y' = g$, because this would violate uniqueness if there were.

5.3   Models of Interacting Species

Comments

Models of the population of interacting species have been popular for some time. These models (particularly the oversimplified models introduced here) don't have much predictive value, but they are guides to thinking about how species interact. Some would call these models caricatures, rather than models. A caricature is a distortion of reality, but the original is still recognizable—and that is just what these models are like in comparison to the real thing. We restrict attention to two interacting species whose model rate equations are autonomous. That means that time variations in the rate functions have been averaged out—a dubious proposition at best, but the modeling process usually begins with oversimplifications. We give a catalogue of possible interactions between two populations, ranging from the now classical Volterra model for predator-prey interactions to assorted models of cooperation, competition, harvesting, and satiation. No deep results here—just some computer simulations of orbits in the population quadrant $x \geq 0$, $y \geq 0$. We have not explicitly scaled the population levels or time; in fact, we don't even mention units for measuring time or populations. But the variables are all scaled down to convenient levels for computation and calculation.

1. **(a).** Each species modeled by the system $x' = (\alpha - by)x$, $y' = (\beta - cx)y + H$ would grow exponentially in isolation (the rate terms αx, βy), but when occupying the same ecological region (or *niche*, to use the technical term) they compete for common resources (the terms $-bxy$ and $-cxy$). The y-species is restocked at rate H.

    (c). The system is $x' = (\alpha + by)x$, $y' = (-\beta + cx - dy)y$. The x-species would grow exponentially in isolation (the rate term αx), while the y-species would decline exponentially (the rate term $-\beta y$) with the decline accelerated by overcrowding (the term $-dy^2$). The coupling terms bxy and cxy indicate that the two species somehow cooperate, the presence of each being favorable for the growth of the other.

    (e). The system is $x' = (\alpha + by)x - H \operatorname{sqw}(t, 25, 1)$, $y' = (\beta + cx - dy)y$. Each species would grow exponentially in isolation (rate terms αx and βy), except that overcrowding implies logistic change for the y-species (the rate term $-dy^2$). When occupying the same ecological niche the two species cooperate, the presence of each being favorable for the other (the terms bxy and cxy). The x-species is seasonally harvested at the rate H, but only for the first 25% of the period 1. This is a model of cooperation (the terms bxy and cxy).

    (g). The system is $x' = (\alpha - bz)x$, $y' = (\beta - my - kz)y$, $z' = (-\gamma + az + cy)z - Hz$. The y-species would change logistically if isolated (the term $\beta y - dy^2$), but the x-species would increase exponentially if isolated (the term dx), and the z-species would decline exponentially if isolated (the term $-\gamma z$). The z-species is a predator on the prey species x and y (the terms $-bzx$ and axz, $-kzy$ and cyz). The z-species is harvested with constant effort harvesting (the term $-Hz$).

3. The dashed lined are the nullclines.

    (a). See Fig 3(a) for orbits of $x' = (5 - x + y)x$, $y' = (10 + x - 5y)y$. Each population tends to a positive equilibrium value as $t \to +\infty$.

    (c). See Fig 3(c). The y-species wins the competition as $x \to 0$ and $y \to 5$ as $t \to \infty$.

Problem 3(a).

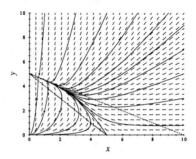

Problem 3(c).

5. The system is $x' = (2 + H - x - 2y)x$, $y' = (2 - 2x - y)y$. The equilibrium point de-

termined by the intersection of the x-nullcline $2 + H - x - 2y = 0$ and the y-nullcline $2 - 2x - y = 0$ has coordinates $x = (2 - H)/3$, $y = 2(1 + H)/3$. This point exits the first quadrant as H increases through the value 2. The other equilibrium points are $(0, 0)$, $(0, 2)$, and $(2 + H, 0)$. For $0 < H < 2$, it seems that we have a model of competitive exclusion (see Fig. 5, Graph 1 where $H = 1$). At $H = 2$, the x-species seems to win the competition (see Fig. 5, Graph 2 where $H = 2$). For $H > 2$, the x-species wins (see Fig. 5, Graph 3 where $H = 3$).

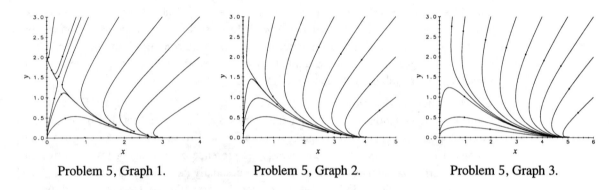

Problem 5, Graph 1.   Problem 5, Graph 2.   Problem 5, Graph 3.

7.   Group project.

5.4 Predator-Prey Models

Comments

This section is devoted to one topic: the Lotka-Volterra models of predator-prey dynamics, and the story of how Volterra came to consider the question. The model isn't very realistic, although (as we point out) it is based on data from real fish populations. Anyway, Volterra derived three laws from the model equations, and all three are presented and discussed. The implicit formula for the population cycles is derived, but the proof that the formula defines a simple closed curve is not at all easy (see Problem 6). As in Section 5.3, we assume that time and populations have been scaled to appropriate levels for ease in calculation and computation.

1.   **(a).** The system is $x' = -x + xy$, $y' = y - xy$. We see that $x(t)$ is the predator population and $y(t)$ is the prey population, as we can tell from the signs of the interaction terms xy in the rate equations. The average predator and prey populations are 1 since the coordinates of the internal equilibrium point are $x = 1$, $y = 1$ (see Theorem 5.4.2). Because $x' < 0$ if $x > 0$, $y = 0$, and $y' > 0$ if $y > 0$, $x = 0$, the orbits turn clockwise about $(1, 1)$ as time increases.

(c). The system is $x' = (-1 + 0.09y)x$, $y' = (5 - x)y$. We see that $x(t)$ is the predator population and $y(t)$ is the prey population at time t (look at the signs of the xy terms). The average predator and prey populations are 5 and 100/9, respectively, because $(5, 100/9)$ is the internal equilibrium point (see Theorem 5.4.2). Because $x' < 0$ if $x > 0$, $y = 0$, and $y' > 0$ if $y > 0$, $x = 0$, the orbits turn clockwise about $(5, 100/9)$ as time increases.

3. The xy-system is $x' = f(x, y) = -ax + bxy$, $y' = g(x, y) = cy - dxy$. Let $x_1 = c/d$ and $y_1 = a/b$ so (x_1, y_1) is the equilibrium point.

(a). To linearize we first find the partial derivatives of $f(x, y)$ and $g(x, y)$ with respect to both x and y. These are $\partial f/\partial x = -a + by$, $\partial f/\partial y = bx$, $\partial g/\partial x = -dy$, and $\partial g/\partial y = c - dx$. So with $x_1 = c/d$ and $y_1 = a/b$, we have $x' = \partial f/\partial x(x_1, y_1)(x - x_1) + \partial f/\partial y(x_1, y_1)(y - y_1) = (-a + b(a/b))(x - c/d) + b(c/d)(y - a/b) = bcy/d - ac/d$ and $y' = \partial g/\partial x(x_1, y_1)(x - x_1) + \partial g/\partial y(x_1, y_1)(y - y_1) = -d(a/b)(x - c/d) + (c - d(c/d))(y - a/b) = -adx/b + ac/b$. The linearized system is $x' = bcy/d - ac/d$, $y' = -adx/b + ac/b$.

(c). See Fig. 3(c), Graph 1 for the graphs of the predator-prey orbits and component graphs of the nonlinear Lotka-Volterra system, $x' = (-1 + y)x$, $y' = (1 - x)y$. Graph 2 shows the orbits and component graphs for the corresponding linearized system. Note from the x-component graphs how the period of the predator-prey cycles of the nonlinear system increases with amplitude, but the period of the linearized cycles is fixed at $2\pi/\sqrt{ac}$ $(= 2\pi$ if $a = c = 1$). The initial points used for the graphs in both figures are $x_0 = 1$, $y_0 = 1, 1.1$, $1.3, 1.5$, and 1.9.

Problem 3(c), Graph 1.

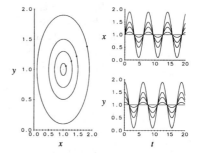

Problem 3(c), Graph 2.

5. **(a).** Figure 5.4.3 graphs the system $x' = -x + xy/10 - Hx$, $y' = y - xy/5 - Hy$ with initial conditions $x(0) = 8$, $y(0) = 15$ for $H = 0, 2/5, 1, 5$. We see two extinction orbits in the graph. When $H_1 = H_2 = 1.0$, the predator species vanishes, but not the prey. When $H_1 = H_2 = 5.0$, the heavy harvesting kills off both species. The other orbits correspond to lower values of $H_1 = H_2$, where neither population becomes extinct.

(c). If $H_2 = c$, then replacing the coefficient d by d^* to avoid confusing it with the "d" of the differential,

$$\frac{dy}{dx} = \frac{d^*xy}{(a + H_1)x - bxy} = \frac{d^*y}{a + H_1 - by}$$

After separating variables, we have

$$\left(\frac{a + H_1 - by}{d^* y}\right) dy = dx$$

So, $(a + H_1)/d^* \cdot (1/y)dy - (b/d^*)dy = dx$. Integrating, $((a + H_1)/d^*) \ln(y/y_0) - (b/d^*)(y - y_0) = x - x_0$; that is,

$$x = x_0 + \frac{a + H_1}{d} \ln \frac{y}{y_0} - \frac{b}{d}(y - y_0)$$

7.   Group project.

5.5   The Possum Plague: A Model in the Making

Comments

The models for the possum/tuberculosis problem are still in the works. So what we present in this section may very well be modified or even discarded within a few years. This material gives a glimpse at a model in the making and offers the reader a chance to play with the parameters of the model and maybe come up with some suggestions. Some may want to read the article by Wake and his colleagues cited in the footnote. We have scaled the variables and the parameters one way, but there are other ways to do the scaling, and the reader may want to try something else. We have included a traditional SIR (susceptible, infective, recovered) model in the problem set. Note that the S and I rate equations of the SIR model in Problem 1 have the same form as the rate equations for a harvested predator-prey system if the harvesting coefficient of the prey has the same magnitude as the natural growth coefficient [see Section 5.4, Problem 5(d)].

1.   **(a).** Since the parameter b represents the reciprocal of the period of infection, the rate equation for the population of recovered is $R' = bI$. The rate equation for the population of susceptibles is $S' = -aSI$ because the spread of the disease depends upon contact between susceptibles and infectives, and so a mass action term such as $-aSI$ is appropriate, where $1/a$ measures the level of exposure of the average susceptible to the average infective. Since we assume that the total population is constant, then $S + I + R =$ constant, which implies that $I' = -(S' + R') = aSI - bI$.

(c). We have $a = 0.003 \cdot (1/6) = 0.0005$. Since the duration of the illness in a person is four days, $b = 1/4 = 0.25$. It is shown in part **(b)** that the epidemic occurs if $S > 2b/a = 2 \times 0.25/0.0005 = 1000$. See Fig. 1(c) for graphs of $I(t)$, where $I(0) = 50$, $S_0 = 500, 1000, 1500$ (solid, long dashes, and short dashes, respectively). Note that an epidemic occurs only when the susceptible population is large enough (in the plotted cases, $S(0) = 1000, 1500$).

(e). Because $dI = [b/(aS) - 1]dS$, the general solution is $I = (b/a)\ln S - S + C$. According to the initial condition, we know that $I = (b/a)\ln(S/S_0) + (S_0 - S) + I_0$. See

Fig. 1(e), Graph 1 where $a = 0.001$, $b = 0.08$. The four curves (from top down) correspond to SI-initial data $(1500, 50)$, $(1250, 35)$, $(1000, 20)$, $(800, 15)$; the time span is $0 \le t \le 50$. The S-axis is a line of equilibrium points, and it seems that each SI-orbit starts (at $t = -\infty$) at a point $(S_1, 0)$ where S_1 is large, and ends (as $t \to +\infty$) at $(S_2, 0)$ where S_2 is small. This is plausible since as a disease runs its course the susceptible pool diminishes and the number of infectives first increases and then diminishes. Fig. 1(e), Graph 1 is deceptive because it suggests that as time goes on the number of susceptibles tends to 0. Fig. 1(e), Graph 2 zooms on the left-hand ends of the SI-orbits, and we see that as $t \to +\infty$ the separate orbits actually approach different points on the S-axis.

Problem 1(c).

Problem 1(e), Graph 1.

Problem 1(e), Graph 2.

3.   Let $S = P - I$, that is, $P = S + I$. Substituting into the possum model, we have $I' = \beta IS - (\alpha + b)I$ and $S' = (a - b)(S + I) - \alpha I - I' = (a - b)S + aI - \beta IS$. Similarly, from (5) with $z = x - y$ (so z is the scaled population of susceptible possums), we have $dy/ds = yz - ry = (z - r)y$, and $dz/ds + dy/ds = c(y + z) - y$. So the z and y rate equations are

$$dz/ds = z(c - y) - (1 - c - r)y \qquad dy/ds = (z - r)y$$

The corresponding equilibria are $(r, cr/(1 - c))$ and $(0, 0)$.

    In Fig. 3, Graph 1 ($c = -1$, $r = 2$), the possums and the disease all disappear as $t \to +\infty$, basically because $c = -1$, and so the possum death rate due to natural causes (not tuberculosis) exceeds the birth rate. In Fig. 3, Graph 2 ($c = 0$, $r = 2$). The natural birth and death rates balance. In this case all points $z > 0$, $y = 0$ are equilibrium points, and every orbit tends to one of these equilibrium points as $y(t)$ diminishes and t tends to $+\infty$. Distinct orbits tend to distinct equilibrium points, although that is far from evident in Fig. 3, Graph 2. See Fig. 3, Graph 3 for a zoom of Fig. 3, Graph 2 that separates orbits near the equilibrium points. In Fig. 3, Graph 4 ($c = 1.5$, $r = 2$) the constant c is so large that the number of infectives increases without bound while the number of susceptible possums remains small. The population orbits in Fig. 3, Graph 5 ($c = 0.25$, $r = 2$) spiral toward the equilibrium point $(2, 2/3)$. The interpretation is that here the disease tends to a fairly low level endemic state. The interpretation of Fig. 3, Graph 6 ($c = 0.05$, $r = 1.1$) is similar to that of Fig. 3, Graph 5.

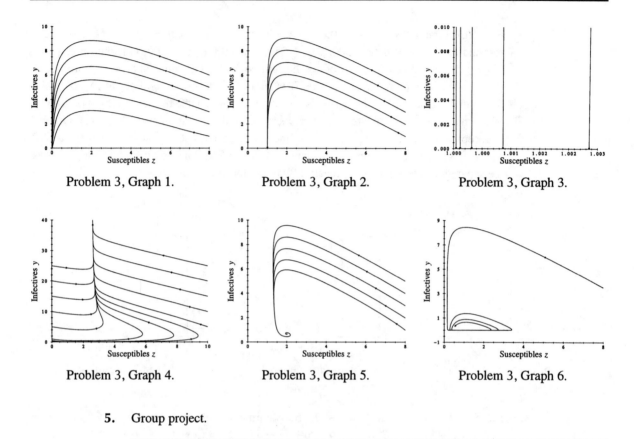

Problem 3, Graph 1.   Problem 3, Graph 2.   Problem 3, Graph 3.

Problem 3, Graph 4.   Problem 3, Graph 5.   Problem 3, Graph 6.

5.   Group project.

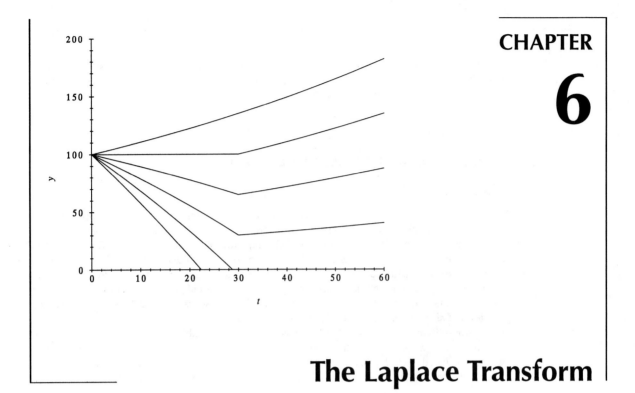

CHAPTER

6

The Laplace Transform

There are differences of opinion about the place of Laplace transforms in an ODE course. Institutions with an engineering program often offer a "baby systems" course which contains a healthy dose of Laplace transforms. This reflects the central role played by the Laplace (and Fourier) transform in the work of practicing engineers. Because of this, some feel that the time in an ODE course can be better spent in doing some state plane analysis, stability, and bifurcations, subjects that engineers may not otherwise see in the classroom. On the other hand, institutions without an engineering program of their own may very well want to put Laplace transforms in their ODE course. This chapter has been included just in case it is needed, but no harm is done in skipping the chapter altogether.

With the above remarks out of the way, we make a few general observations about the contents of Chapter 6. The chapter is quite demanding technically, requiring, as it does, a great deal of algebraic pyrotechnics and much skill in using tables and the calculus of transforms. For convenience in calculating Laplace transforms, we have put a table giving the transforms of commonly used functions at the end of the chapter. Partial fractions play a big role in using Laplace transforms to solve IVPs, and the basic expansions are given in Table 6.2.1.

In using Laplace transforms to solve IVPs, the attempt will fail unless the ODEs are linear, and will be very difficult even if linear if the coefficients are not constant. The big advantage of transforms lies in their ability to handle discontinuous driving terms such as square waves or sawtooth waves and in the simple way the transform involves initial data right from the start of the solution process. The disadvantage (other than the restriction to linear ODEs with constant coefficients) is that the formula for the inverse transform is beyond the scope of this book. And so we must be satisfied with inversion by pattern matching from tables and by the calculus of transforms. It is a remarkable fact, however, that Laplace transforms can be used to solve constant-coefficient linear operator equations other than linear ODEs. An example of the transform technique to difference equations is given in Problem 10, Section 6.2, and to an unusual car-following model that involves differential-delay equations in Sections 6.1 and 6.3.

6.1 Introduction to the Laplace Transform

Comments

The purpose of this short section is to give a quick introduction to the transform, the ability to calculate the transform of some elementary functions, and a glimpse at applications of Laplace transforms to solving IVPs. The curious car-following process is modeled here, but the model equations are only solved in Section 6.3. The inverse transform is introduced (but not by formula). We note that although distinct continuous functions have distinct transforms, two different discontinuous functions may have the same transform. Logically, we should introduce equivalence classes of piecewise-continuous functions at this point so that the inverse transform L^{-1} is well-defined, but that level of abstraction seems out of place at the introductory level. How do you set about finding transforms—brain only, calculus book, tables of integrals, or tables of Laplace transforms? Our preference is to work out a few transforms from scratch, but after that to use tables, which is why we have included a set of transform tables at the end of the chapter. It is not a good idea to spend so much time working out integrals that you lose sight of what the transform is all about.

---

1. The transform $L[f](s) = \int_0^\infty e^{-st} f(t)\, dt$ may be computed by consulting tables of integrals or directly through integration by parts. The details of **(a)** are given, but answers only are listed for the other parts.

   (a).

   $$L[3t - 5] = \int_0^\infty e^{-st}(3t - 5)\, dt = \left[3\frac{e^{-st}}{s^2}(-st - 1) + \frac{5}{s}e^{-st} \right]\Big|_{t=0}^{t=\infty} = \frac{3}{s^2} - \frac{5}{s}$$

   since $\lim_{t \to +\infty} e^{-st} = 0$ and $\lim_{t \to +\infty} te^{-st} = 0$ if (and only if) $s > 0$. And so, $0 < s < \infty$ is the largest interval on which the transform is defined.

   (c). $L[t^n] = n!/s^{n+1}$, $s > 0$ (using tables or repeated integration by parts).

   (e). $L[te^{at}] = 1/(s - a)^2$, $s > a$ (using tables or integration by parts).

3. Apply the operator L to each side of the ODE and use the formula $L[y'](s) = sL[y](s) - y(0)$.

   (a). The IVP is $y' + 2y = 0$, $y(0) = 1$. Applying L, we have that $sL[y] - 1 + 2L[y] = 0$. We have $L[y] = 1/(s + 2)$. Since $L[e^{-2t}] = 1/(s + 2)$ (see Example 6.1.3 or II.3 in the tables), we have that $y = e^{-2t}$.

   (c). The IVP is $y' + 2y = e^{it}$, $y(0) = 0$. Applying L, we have $sL[y] + 2L[y] = 1/(s - i)$, where II.3 in the tables has been used with $a = i$. So, $L[y] = 1/[(s + 2)(s - i)]$. By II.9 with $a = -2$, and $b = i$, we have $y = (e^{it} - e^{-2t})/(2 + i)$.

6.2 Calculus of the Transform

Comments

Laplace transforms are a great tool for solving IVPs, but, as with any tool, you have to learn the operational rules first—and there are a lot of them. We have put the most important of the rules in this section—how do you transform a derivative, an integral, or an on-off function? Going backward, how do you inverse transform a quotient of polynomials? An important part of the answer to the last question is the algebraic technique of partial fractions (see Table 6.2.1). The Shifting Theorem (the three formulas in Theorem 6.2.4) are essential when dealing with functions where the time scale (or the s-variable scale) has been shifted. You may want to use a CAS to avoid what are often lengthy and exacting algebraic manipulations.

1. The transforms can be obtained by straightforward integration, integration by parts, tables of integrals, or the tables at the chapter's end. For brevity, we refer to the tables or to formulas derived in this section. The basic partial fraction expansions are given in Table 6.2.1; every expansion needed here can be reduced to one in the table.

   (a). Since $\sinh at = (e^{at} - e^{-at})/2$, $L[\sinh at] = a/(s^2 - a^2)$ by II.7 (or II.3).

   (c). By Eq. (11), $L[t^2 e^{at}] = L[t^2](s - a) = 2/(s - a)^3$, where we have used II.1.

   (e). Using Eq. (11) and Eq. (4), $L[te^{2t} f(t)] = L[tf(t)e^{2t}] = L[tf(t)](s - 2) = -\phi'(s - 2)$, where $L[f](s)$ is denoted by $\phi(s)$.

   (g). $L[(t + 1)\operatorname{step}(t - 1)] = L[(t + 2 - 1)\operatorname{step}(t - 1)] = e^{-s} L[t + 2] = e^{-s}(1/s^2 + 2/s)$ by Eq. (12) and II.3.

   (i). $L\left[e^{at}[\operatorname{step}(t - 1) - \operatorname{step}(t - 2)]\right] = e^{-s} L[e^{a(t+1)}] - e^{-2s} L[e^{a(t+2)}] = e^{-(s-a)}/(s - a) - e^{-2(s-a)}/(s - a)$, where we have used Eq. (13).

3. Apply the operator L to each side of the ODE in the given IVP, and use formula (2) with $n = 2$. See Table 6.2.1 for partial fraction expansions.

   (a). The IVP is $y'' - y' - 6y = 0$, $y(0) = 1$, $y'(0) = -1$, so $s^2 L[y] - s + 1 - sL[y] + 1 - 6L[y] = 0$. Using partial fractions

$$L[y] = \frac{s - 2}{s^2 - s - 6} = \frac{s - 2}{(s - 3)(s + 2)} = \frac{1}{5(s - 3)} + \frac{4}{5(s + 2)}$$

   So, $y(t) = (e^{3t} + 4e^{-2t})/5$ by II.3.

   (c). The IVP is $y'' - 2y' + 2y = 0$, $y(0) = 0$, $y'(0) = 1$, so $s^2 L[y] - 1 - 2sL[y] + 2L[y] = 0$:

$$L[y] = \frac{1}{s^2 - 2s + 2} = \frac{1}{(s - 1)^2 + 1}$$

   By Eq. (11) and II.5, $y(t) = e^t \sin t$.

   (e). The IVP is $y'' - 2y' + y = \operatorname{step}(t - 1)$, $y(0) = 1$, $y'(0) = 0$, so $s^2 L[y] - s - 2sL[y] +$

$2 + L[y] = L[\text{step}(t-1)] = e^{-s}/s$ by (14). Using partial fractions, we have

$$L[y] = \frac{s-2}{s^2 - 2s + 1} + \frac{e^{-s}}{s(s^2 - 2s + 1)} = \frac{1}{s-1} - \frac{1}{(s-1)^2} + e^{-s}\left(\frac{1}{s} - \frac{1}{s-1} + \frac{1}{(s-1)^2}\right)$$

Then using II.3 and (14), we have

$$y = e^t(1-t) + \text{step}(t-1)[1 - e^{t-1} + (t-1)e^{t-1}] = e^t(1-t) + \text{step}(t-1)[1 + (t-2)e^{t-1}]$$

5. Since

$$\text{step}(t-a) = \begin{cases} 1, & t \geq a \\ 0, & t < a \end{cases} \quad \text{and} \quad \text{step}(t-b) = \begin{cases} 1, & t \geq b \\ 0, & t < b \end{cases}$$

and $b > a$, we have $\text{step}(t-a) - \text{step}(t-b) = 1$ for $a \leq t < b$, and $= 0$ for all $t < a$ or $t \geq b$. This is a square pulse over the interval (a, b), and is denoted by the engineering function $\text{sqp}(t-a, b-a)$. Using the above, $L[\text{sqp}(t-1, 3)] = L[\text{step}(t-1) - \text{step}(t-4)]$, which, using II.19, equals $(e^{-s} - e^{-4s})/s$.

7. The answer to part **(a)** is obtained from **(c)** by setting the resistance $R = 0$. The answer has the same form as in **(c)** because the roots r_1 and r_2 of the characteristic polynomial $s^2 + Rs/L + 1/LC$ are given symbolically.

    (c). Applying the transform to each side of the ODE, $Lq'' + Rq' + \frac{1}{C}q = E_0\,\text{step}(t-a)$, rearranging terms, and using (14), we obtain

$$L[q] = \frac{E_0}{Ls^2 + Rs + 1/C}\frac{e^{-as}}{s} = \frac{E_0/L}{(s-r_1)(s-r_2)s}e^{-as}$$

where r_1 and r_2 are assumed to be distinct roots of $s^2 + Rs/L + 1/(LC)$, but they need not be real. Since we have the partial fraction decomposition

$$\frac{1}{(s-r_1)(s-r_2)s} = \frac{1}{r_1 r_2(r_1 - r_2)}\left[\frac{r_2}{s-r_1} - \frac{r_1}{s-r_2} + \frac{r_1 - r_2}{s}\right]$$

we can use II.3 and formula (12) and the fact that $r_1 r_2 = 1/LC$ to obtain

$$q_a(t) = \text{step}(t-a)CE_0\left[1 + \frac{r_2}{r_1 - r_2}e^{r_1(t-a)} + \frac{r_1}{r_2 - r_1}e^{r_2(t-a)}\right]$$

where $q_a(t)$ is the charge at time $t \geq 0$ if the switch is turned on at time $t = a$.

9. **(a).** The IVP is $x_1' = 3x_1 - 2x_2 + t$, $x_2' = 5x_1 - 3x_2 + 5$, $x_1(0) = x_2(0) = 0$. Using the initial data and applying the Laplace transform to each term in each equation, we have that $sL[x_1] = 3L[x_1] - 2L[x_2] + s^{-2}$, $sL[x_2] = 5L[x_1] - 3L[x_2] + 5s^{-1}$. Solving these equations for the transforms $L[x_1]$ and $L[x_2]$, we have (using a partial fractions decomposition)

$$L[x_1] = \frac{3 - 9s}{s^2(s^2 + 1)} = \frac{-9}{s} + \frac{3}{s^2} + \frac{9s - 3}{s^2 + 1}$$

$$L[x_2] = \frac{5s^2 - 15s + 5}{s^2(s^2 + 1)} = \frac{-15}{s} + \frac{5}{s^2} + \frac{15s}{s^2 + 1}$$

Using the Laplace transform tables, we can invert the transforms to obtain $x_1(t) = -9 + 3t + 9\cos t - 3\sin t$, $x_2(t) = -15 + 5t + 15\cos t$.

6.3 Applications of the Transform: Car Following

Comments

The main topic of this section is the Laplace transform of a periodic function. Of course, we have already shown how to transform sinusoids, but now we derive a formula for the transform of any periodic function. That formula is then applied to an LC circuit modeled by an ODE in the current $I(t)$ with a square wave driving function and to the car-following model of Section 6.1. The details are all worked out, and they are far from trivial since geometric expansions and partial fractions are required. Partial fractions formulas were introduced in Table 6.2.1. As is usually the case when working with transforms, there is a lot of algebra. The geometric series expansion $(1-x)^{-1} = 1 + x + x^2 + \cdots + x^n + \cdots$ for $|x| < 1$ is used throughout the section and its problems. It is valid for any x whose magnitude is less than 1; for example, it is valid for $x = e^{-s}$, $s > 0$.

Traffic models are widely used these days to predict what will happen in congested areas when certain conditions occur. As always when human factors are involved, the models are more suggestive than exact. Nevertheless, they do give a way of thinking about traffic flow, and they do include certain kinds of human behavior that most of us can immediately recognize. The modeling outlined in Section 6.1 and in this section is straightforward. However, solving the model equations by using transform techniques is anything but straightforward. On the other hand, formulas (22)–(24) show that the solution is not that hard to understand and use.

1. Since a periodic function appears as a factor in each case, formula (2) may be used at some point in the calculation. On the other hand, it is often simpler to use integration tables (or integration by parts) or transform formulas from Sections 6.1 or 6.2 if the periodic functions involve sines and cosines. We use both approaches for **(a)**, but give answers only for the rest.

(a). $L[e^t \sin t](s) = L[\sin t](s-1)$ by (11) of the Shifting Theorem (Theorem 6.2.4). By formula (2) and integral tables (or a double integration by parts)

$$L[\sin t](s) = \frac{1}{1 - e^{-2\pi s}} \int_0^{2\pi} e^{-st} \sin t\, dt = \frac{1}{1 - e^{-2\pi s}} \left[\frac{1 - e^{-2\pi s}}{1 + s^2} \right] = \frac{1}{1 + s^2}$$

So, $L[e^t \sin t] = 1/(1 + (s-1)^2)$. Alternatively, by integral tables (or a double integration by parts) we have that

$$L[e^t \sin t](s) = \int_0^\infty e^{-st} e^t \sin t\, dt = \int_0^\infty e^{-(s-1)t} \sin t\, dt$$

$$= \frac{1}{1 + (s-1)^2} \left[e^{-(s-1)t}(-(s-1)\sin t - \cos t) \right]_{t=0}^{t=\infty} = \frac{1}{1 + (s-1)^2}$$

(c). Using the result [from part **(b)**] that $L[\sin^2 t](s) = 2/[s(s^2+4)]$, we have

$$L[\cos^2 t](s) = L[1 - \sin^2 t](s) = \frac{1}{s} - \frac{2}{s(s^2+4)} = \frac{s^2+2}{s(s^2+4)}$$

(e). By formula (4) in Section 6.2 [and part **(d)**] $L[t \sin at](s) = 2as/(a^2+s^2)^2$,

$$L[t^2 \sin at] = -\frac{d}{ds} L[t \sin at] = -\frac{d}{ds}\left[\frac{2as}{(a^2+s^2)^2}\right] = -\frac{2a(3s^2-a^2)}{(a^2+s^2)^3}$$

(g). Using a trigonometric identity, we have

$$L[e^{-3t}\cos(2t+\pi/4)](s) = \frac{1}{\sqrt{2}} L[e^{-3t}\cos(2t)](s) - \frac{1}{\sqrt{2}} L[e^{-3t}\sin(2t)](s)$$

$$= \frac{1}{\sqrt{2}} L[\cos(2t)](s+3) - \frac{1}{\sqrt{2}} L[\sin(2t)](s+3)$$

by (11) of the Shifting Theorem (Theorem 6.2.4). And so by Example 6.2.1,

$$L[e^{-3t}\cos(2t+\pi/4)] = \frac{1}{\sqrt{2}}\left[\frac{s+3}{(s+3)^2+4} - \frac{2}{(s+3)^2+4}\right] = \frac{1}{\sqrt{2}}\frac{s+1}{(s+3)^2+4}$$

3. The partial fraction decompositions and the answers $f(t)$ are listed, but the detailed calculations are omitted. The partial fraction decompositions follow from Table 6.2.1.

(a).

$$\frac{1}{s(s+1)} = \frac{1}{s} - \frac{1}{s+1}, \qquad f(t) = 1 - e^{-t}$$

(c).

$$\frac{1}{(s-a)(s-b)} = \frac{1}{a-b}\left[\frac{1}{s-a} - \frac{1}{s-b}\right], \qquad f(t) = \frac{1}{a-b}[e^{at} - e^{bt}]$$

(e).

$$\frac{3s+1}{(s^2+2s+2)(s-1)} = -\frac{1}{5}\cdot\frac{4s-3}{(s+1)^2+1} + \frac{4}{5}\cdot\frac{1}{s-1}$$

$$f(t) = -\frac{4}{5}e^{-t}\cos t + \frac{3}{5}e^{-t}\sin t + \frac{4}{5}e^{t}$$

5. The three graphs in each case show (from top down) the square wave input and the component graphs of I and I'. The IVP is $I'' + \omega^2 I = 4\,\mathrm{sqw}(t, 50, T) - 2$, $I(0) = 0$, $I'(0) = 0$.

(a). ($\omega^2 = 1$, $T = 5\pi$) The natural period is 2π since $I'' + I = 0$ has solutions $\sin t$ and $\cos t$ of period 2π. The period of the square wave input is 5π, so the common period is 10π. One expects to see a periodic response of period $10\pi \approx 31.4$, and it is visible in the graphs of I and I' vs. t in Fig. 5(a).

(c). ($\omega^2 = 1$, $T = 2\pi$) The natural period and the period of the square wave input are both 2π. One might expect to see resonance, i.e., a response of growing amplitude (see Section 4.2). The graphs of Fig. 5(c) show that this is the case.

Problem 5(a).   Problem 5(c).

7. **(a).** The IVP is $y' + y = \text{sqw}(t, 50, T)$, $y(0) = 0$. $L[y' + y] = sL[y] + L[y] = L[\text{sqw}(t, 50, T)]$. So,

$$L[y] = \frac{1}{s+1} \cdot \frac{1}{s}\left[1 + \sum_{n=1}^{\infty}(-1)^n e^{-nTs/2}\right]$$

where we have used the expansion in Problem 6 of the square wave with $T = 2a$. Since

$$\frac{1}{(s+1)s} = \frac{1}{s} - \frac{1}{s+1}$$

we have

$$L[y] = \frac{1}{s} - \frac{1}{s+1} + \sum_{n=1}^{\infty}(-1)^n e^{-nTs/2}\left[\frac{1}{s} - \frac{1}{s+1}\right]$$

Inverting, we have

$$y = L^{-1}\left[\frac{1}{s}\right] - L^{-1}\left[\frac{1}{s+1}\right] + \sum_{n=1}^{\infty}(-1)^n L^{-1}\left\{e^{-nTs/2}\left[\frac{1}{s} - \frac{1}{s+1}\right]\right\}$$

$$= 1 - e^{-t} + \sum_{n=1}^{\infty}(-1)^n \text{step}(t - nT/2)[1 - e^{-(t-nT/2)}]$$

where we have used formula (12) in Theorem 6.2.4. See Graph 1 where the period of the driving term is $T = 30$, and Graph 2 for $T = 1$. In each case, the response tends to a forced oscillation of period T. Graph 1 and Graph 2 show (from top down) the square wave input and the output $y(t)$.

If the ODE models a communication channel that receives square waves of various periods as inputs, then the long period (i.e., low frequency) input comes out of the channel [as $y(t)$] with much less distortion and amplitude loss (Graph 1 where $T = 30$) than the short period and high-frequency input (Graph 2 where $T = 1$).

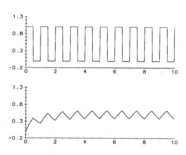

Problem 7(a), Graph 1.   Problem 7(a), Graph 2.

9.   **(a).** Using the expansion of a triangular wave given in Problem 8, we see that

$$\mathcal{L}[y' + y] = s\mathcal{L}[y] + \mathcal{L}[y] = \mathcal{L}[\text{trw}(t, 100, T)]$$

$$= \frac{2}{Ts^2}\left[1 + 2\sum_{n=1}^{\infty}(-1)^n e^{-nTs/2}\right]$$

and so

$$\mathcal{L}[y] = \frac{2}{T(s+1)s^2}\left[1 + 2\sum_{n=1}^{\infty}(-1)^n e^{-nTs/2}\right]$$

Since

$$\frac{1}{(s+1)s^2} = \frac{1}{s+1} - \frac{1}{s} + \frac{1}{s^2}$$

we have

$$\mathcal{L}[y] = \frac{2}{T}\left(\frac{1}{s+1} - \frac{1}{s} + \frac{1}{s^2}\right) + \frac{4}{T}\sum_{n=1}^{\infty}(-1)^n e^{-nTs/2}\left(\frac{1}{s+1} - \frac{1}{s} + \frac{1}{s^2}\right)$$

Inverting, we have

$$y = \frac{2}{T}\left(e^{-t} - 1 + t\right) + \frac{4}{T}\sum_{n=1}^{\infty}(-1)^n \text{step}(t - \frac{nT}{2})\left[e^{-(t-nT/2)} - 1 + (t - \frac{nT}{2})^2\right]$$

See Graph 1 for the input and response when $T = 30$ and Graph 2 for the case $T = 1$. This is a model for a low-pass communication channel [see also Problem 7(a)]. The channel transmits the long-period (low-frequency) triangular wave input without much distortion, but badly distorts the short-period (high-frequency) triangular wave input. The two graphs in each case show (from top down) the triangular wave input and the output $y(t)$.

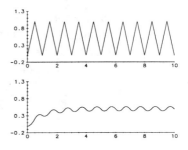

Problem 9(a), Graph 1.　　　　　　　Problem 9(a), Graph 2.

11. (a), (c). Using the last inequality in (25), we see that a collision occurs in every case, because $\lambda T > 12 + 5/T^2$ for the values $T = 1$, $\lambda T = 20$[part (a)] and $T = 2$, $\lambda T = 14$[part (c)].

13. (a). The model is much like the one in system (13) in the text, but separation of cars, rather than the difference in velocities, is the stimulus. If the acceleration of the driver $j + 1$ is α_{j+1} and the response time and sensitivity are T seconds and λ (seconds) $^{-1}$, respectively, then $\alpha_{j+1}(t + T) = \lambda[y_j(t) - y_{j+1}(t)]$, $j \geq 1$. As before, $v_1(t) = f(t)$ for the lead car, so $y_1(t) = \int_0^t f(s)\,ds$.

(c). We have that $y_1(t) = \alpha_0 t^2/2$, $y_2''(t + T) = \lambda[\alpha_0 t^2/2 - y_2(t)]$. Since $\mathcal{L}[y'(t + T)] = e^{Ts}\mathcal{L}[y(t)]$ [see formula (17) in the text], we have that $\mathcal{L}[\alpha_2(t + T)] = \mathcal{L}[y_2''(t + T)] = e^{Ts}\mathcal{L}[y_2'(t)] = e^{Ts}\{s\mathcal{L}[y_2] - y_2(0)\} = e^{Ts}\{s\mathcal{L}[y_2] + 5\}$. On the other hand, from part (a), $\mathcal{L}[\alpha_2(t + T)] = \lambda\mathcal{L}[y_1(t) - y_2(t)] = \lambda\mathcal{L}[\alpha_0 t^2/2 - y_2] = \alpha_0\lambda/s^3 - \lambda\mathcal{L}[y_2]$. So,

$$\mathcal{L}[y_2] = \frac{\alpha_0\lambda s^{-3} - 5e^{Ts}}{se^{Ts} + \lambda} = \frac{1}{se^{Ts}}(\alpha_0\lambda s^{-3} - 5e^{Ts})\left[1 + \frac{\lambda}{s}e^{-Ts}\right]^{-1}$$

Expanding the last factor in a geometric series, we have that

$$\mathcal{L}[y_2] = \frac{1}{se^{Ts}}\left(\alpha_0\lambda s^{-3} - 5e^{Ts}\right)\left(1 - \frac{\lambda}{s}e^{-Ts} + \frac{\lambda^2}{s^2}e^{-2Ts} - \cdots\right)$$

$$= \left(\frac{-5}{s} + \frac{\alpha_0\lambda e^{-Ts}}{s^4}\right)\left(1 - \frac{\lambda}{s}e^{-Ts} + \frac{\lambda^2}{s^2}e^{-2Ts} - \cdots\right)$$

$$= \frac{-5}{s} + \left(\frac{\alpha_0\lambda}{s^4} + \frac{5\lambda}{s^2}\right)e^{-Ts} + \left(\frac{-5\lambda^2}{s^3} - \frac{\alpha_0\lambda^2}{s^5}\right)e^{-2Ts} + \cdots$$

So

$$y_2(t) = -5 + \lambda H(t - T)[\frac{\alpha_0}{6}(t - T)^3 + 5(t - T)]$$

$$- \lambda^2 H(t - 2T)\left[\frac{\alpha_0}{24}(t - T)^4 + \frac{5}{2}(t - T)^2\right] + \cdots$$

The first and second "point" cars collide if $y_2(t) = y_1(t)$. The values of α_0, λ, T determine whether the first two cars collide, and if they do, when the collision occurs. For example, let $\alpha_0 = 6$ ft/sec^2 and so $y_1(t) = 3t^2$. Suppose that $T = 2$ sec. Let's find a value of λ such that there is a collision at some time t, $2 < t < 4$. We have that $y_1(2) - y_2(2) = 12 + 5 = 17 > 0$, while $y_1(4) - y_2(4) = 48 - [-5 + \lambda(8 + 10)] = 53 - 18\lambda$. If the sensitivity λ exceeds $53/18 \approx 2.9$ sec^{-1}, there is a collision at some time $\bar{t}$, $2 < \bar{t} < 4$ since $y_1(t) - y_2(t)$ reverses sign on the interval $[2, 4]$.

6.4 Convolution

Comments

The Convolution Theorem says that the Laplace transform of a product, $L[fg]$ is equal to the convolution product of $L[f]$ adnd $L[g]$ which is denoted by $L[f] * L[g]$. So the transform of a product is *not* the product of the transforms. At the same time, the convolution product gives a practical way to construct the solution of $P(D)[y] = f$ (with zero initial data) for any constant coefficient operator $P(D)$ and integrable input f. The solution is $y = g * f$, where $g(t) = L^{-1}[1/P(s)]$. The convolution product is a neat idea and new to many readers. The more theoretically inclined like to prove its various properties, and all should appreciate its power in representing solutions of $P(D)[y] = f$ (with zero initial data). Because the approach in this section is to focus on the idea of the convolution, we have deliberately kept calculations to a minimum in the text and in the problems. This will come as a welcome relief after all of the laborious computations of the previous sections.

1. **(a).** Using the Convolution Theorem with $f(t) = t$, $g(t) = t$, we see that

$$L\left[\int_0^t (t - u)u\,du\right] = L[t]L[t] = 1/s^4$$

(c). Since $t^2 - 2tu + u^2 = (t - u)^2$, let $f(t) = t^2$, $g(t) = 1$. The transform of the integral is (by the Convolution Theorem)

$$L\left[\int_0^t (t^2 - 2tu + u^2)\,du\right] = L[t^2]L[1] = 2/s^4$$

3. In each case the Convolution Theorem can be used to write $y(t)$ as an integral by arranging $L[y](s)$ as an appropriate product of functions of s. The transforms are given below along with the final solutions in integral form.

(a). The IVP is $y'' + y = t\,\text{step}(t - 1)$, $y(0) = y'(0) = 0$.

$$L[y] = \frac{1}{s^2 + 1} \cdot \frac{e^{-s}}{s^2}; \quad y(t) = \int_0^t \sin(t - u)[u\,\text{step}(u - 1)]\,du$$

(c). The IVP is $2y'' + y' - y = f(t)$, $y(0) = y'(0) = 0$.

$$L[y] = \frac{1}{3}\left[\frac{1}{s - 1/2} - \frac{1}{s + 1}\right]L[f]; \quad y(t) = \frac{1}{3}\int_0^t f(t - u)\left[e^{u/2} - e^{-u}\right]du$$

(e). The IVP is $y'' + 4y = \begin{cases} 0, & 0 \le t \le 1 \\ t - 1, & 1 \le t \le 2 \\ 1, & t > 2 \end{cases}$, $\quad y(0) = y'(0) = 0.$

$$\mathcal{L}[y] = \frac{1}{s^2 + 4}\mathcal{L}\big[(t - 1)\,\text{step}(t - 1) - (t - 2)\,\text{step}(t - 2)\big] = \frac{1}{s^2 + 4} \cdot \frac{1}{s^2}\left(e^{-s} - e^{-2s}\right)$$

$$y(t) = \frac{1}{2}\int_0^t \sin 2(t - u)\big[(u - 1)\,\text{step}(u - 1) - (u - 2)\,\text{step}(u - 2)\big]\,du$$

5. In each case we write the answer as a convolution of the input function on the right side of the ODE with $\mathcal{L}^{-1}[1/P(s)]$, where $P(D)$ is the operator of the ODE. Use Theorem 6.4.3.

   (a). The IVP is $y'' + 6y' + 13y = f(t)$, $y(0) = y'(0) = 0$. So, $y = [(e^{-3t}\sin 2t)/2] * f(t)$, since $\mathcal{L}^{-1}[(s^2 + 6s + 13)^{-1}] = (1/2)e^{-3t}\sin 2t$.

   (c). The IVP is $y''' + y = t$, $y(0) = y'(0) = y''(0) = 0$. So, the solution of this IVP is

   $$y = \left[e^{-t}/3 - e^{t/2}\left(\frac{1}{3}\cos\frac{\sqrt{3}t}{2} - \frac{1}{\sqrt{3}}\sin\frac{\sqrt{3}t}{2}\right)\right] * t$$

   since

   $$\mathcal{L}^{-1}[1/(s^3 + 1)] = e^{-t}/3 - e^{t/2}\left(\frac{1}{3}\cos\frac{\sqrt{3}t}{2} - \frac{1}{\sqrt{3}}\sin\frac{\sqrt{3}t}{2}\right).$$

7. We have that $e^{at}(f * g)(t) = e^{at}\int_0^t f(t - u)g(u)\,du$, while

   $$e^{at}f(t) * e^{at}g(t) = \int_0^t e^{a(t-u)}f(t - u)e^{au}g(u)\,du = e^{at}\int_0^t f(t - u)g(u)\,du$$

   so $e^{at}(f * g)(t) = e^{at}f(t) * e^{at}g(t)$.

6.5 Convolution and the Delta Function

Comments

The elusive delta function finally makes its appearance, and the we are faced with a hard choice: "Define" the delta function in a slightly shady way as we do, just "wave one's hands," or do enough with distributions to give the whole thing a mathematically respectable aura. At various times in various courses, we have used each of these approaches—and with varying success. The approach we take in this section is as good (or as bad) as any other we have tried. The essential properties of the delta function are contained in Theorem 6.5.1. The model of a vibrating spring that is subjected to an impulsive force helps to flesh out the idea of the delta function.

Experiment

If you are willing to risk the possibility of making a fool of yourself, here is a marvelous opportunity for a demonstration: ask a physicist to give you a dependable spring with a known spring constant

ω^2. Then hang it up, put a weight on it, start it oscillating and then suddenly hit it and bring it to a stop. Or better yet, organize a team to do it, or do a computer simulation (see Example 6.5.2).

---

1. Apply the Laplace transform operator $\mathcal{L}$ to each side of the ODEs, using the property of the delta function that $\mathcal{L}[\delta(t-u)](s) = e^{-us}$. Also, the Shifting Theorem (Theorem 6.2.4) is sometimes used. In every case the initial values are $y(0) = y'(0) = 0$.

(a). The ODE is $y'' + 2y' + 2y = \delta(t-\pi)$. We have that

$$\mathcal{L}[y](s) = \frac{e^{-\pi s}}{s^2 + 2s + 2} = \frac{e^{-\pi s}}{(s+1)^2 + 1}$$

and so $y(t) = \text{step}(t-\pi)f(t-\pi)$, where $\mathcal{L}[f](s) = 1/[(s+1)^2 + 1]$, and $f(t) = e^{-t}\sin t$; $y(t) = \text{step}(t-\pi)e^{-(t-\pi)}\sin(t-\pi) = -\text{step}(t-\pi)e^{-(t-\pi)}\sin t$.

(c). The ODE is $y'' + 3y' + 2y = \sin t + \delta(t-\pi)$. Here, we have

$$\mathcal{L}[y](s) = \frac{1}{(s^2+1)(s+1)(s+2)} + \frac{e^{-\pi s}}{(s+1)(s+2)}$$

After using partial fractions, we have

$$y(t) = [-3\cos t + \sin t + 5e^{-t} - 2e^{-2t}]/5 + \text{step}(t-\pi)[e^{-(t-\pi)} - e^{-2(t-\pi)}]$$

(e). The ODE is $y'' + y = e^t + \delta(t-1)$. In this case

$$\mathcal{L}[y] = \frac{1}{(s^2+1)(s-1)} + \frac{e^{-s}}{s^2+1}$$

After using partial fractions, we have $y = [e^t - \cos t - \sin t]/2 + \text{step}(t-1)\sin(t-1)$.

3. The charge IVP is

$$Lq'' + \frac{1}{C}q = F_0 + 2F_0\delta(t-1), \qquad q(0) = q'(0) = 0$$

Applying $\mathcal{L}$, we have that

$$\mathcal{L}[q](s) = \frac{F_0/L}{s(s^2 + 1/LC)} + \frac{2F_0 e^{-s}/L}{s^2 + 1/LC}$$

Using partial fractions and then inverting, we have

$$q(t) = CF_0\left[1 - \cos\left(\frac{t}{\sqrt{LC}}\right)\right] + 2F_0\,\text{step}(t-1)\sqrt{C/L}\sin\left(\frac{t-1}{\sqrt{LC}}\right)$$

See Fig. 3 for the solution curve where $L = 1$, $C = 1$, $F_0 = 10$. The burst of voltage at $t = 1$ introduces energy into the circuit, and the charge then oscillates with higher amplitude sinusoids.

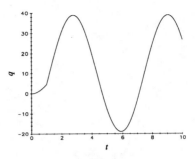

Problem 3.

5. Let's show that $\delta(at) = (1/|a|)\delta(t)$, $a \neq 0$. Let f be any function in $\mathbf{E}_1$. We will make the substitution $v = au$ to obtain

$$\int_{-\infty}^{\infty} \delta(at - au) f(u)du = \frac{1}{a} \int_{-\infty}^{\infty} \delta(at - v) f(v/a)dv$$

$$= \frac{1}{a} f(\frac{at}{a}) = \frac{1}{a} f(t) \text{ if } a > 0$$

if $a > 0$. If $a < 0$, then we have

$$\frac{1}{a} \int_{\infty}^{-\infty} (\cdot) = -\frac{1}{a} \int_{-\infty}^{\infty} (\cdot) = \frac{1}{|a|} \int_{\infty}^{-\infty} (\cdot)$$

Since the above holds for all functions f, we have that $\delta(at) = |a|^{-1}\delta(t)$ if $a \neq 0$.

CHAPTER

7

Linear Systems of Differential Equations

It is hard to judge how much linear algebra readers of this text know before beginning a serious study of linear differential systems. There is a growing trend in many engineering curricula to drop linear algebra as a required course, expecting people to pick it up along the way in other courses. So some will have had a lot of exposure to linear algebra and a growing number may not. But our experience is that even people who have had a linear algebra course are often quite rusty when they have to use it. For these reasons, we have included several sections on matrices, systems of linear (algebraic) equations, determinants, eigenvalues, and eigenvectors. The notion of a linear space is gently described so that we can talk about linear independence. We have tried to make the matrix material as self-contained as possible, and there are examples of every kind of computation needed for the chapter. We recommend that these concepts be thoroughly reviewed by the reader.

After the sections on matrix algebra and eigenvalues, we treat undriven linear systems with constant coefficients, $x' = Ax$. The operator L with action $L[x] = x' - Ax$ is introduced to simplify the presentations of ideas and some of the computations. A gallery of orbital portraits of planar systems with constant coefficients is displayed. This is a place where people can have some fun with their solvers. Basically we just explore, via computer simulation, the behavior of the orbits of planar systems. We also introduce second-order linear systems and solve a problem of coupled springs introduced in Section 5.1. We introduce the exponential matrix as a convenient way of finding a formula for the solution of the IVP, $x' = Ax$, $x(0) = x^0$. The exponential matrix comes in handy when the method of Variation of Parameters is used to solve the IVP, $x' = Ax + F(t)$, $x(0) = 0$. An interesting feature of this chapter is our treatment of the steady-state solution where F is a constant or periodic driving term. Along with this we present some useful tests for determining if all the eigenvalues of the system matrix A have negative real parts without actually finding the eigenvalues: the Routh Test, the Gerschgorin Disks, plus other tests. The solutions of a linear system whose system matrix A has only eigenvalues with negative real parts have some very appealing features. Finally, we extend everything to the general linear system $x' = A(t)x + F(t)$.

There are three "big" models in this chapter. Lead in the human body is the subject of the entire first section and half of the next-to-the-last section. Coupled spring models are solved in Section 7.7. The low-pass filter (also used as a noise filter) is introduced and extensively studied in Section 7.10. The lead and filter models offer nice examples of the sensitivity of linear systems to changing data.

7.1 Tracking Lead Through the Body

Comments

This section is devoted to a single model as the title indicates. It provides a nice example of applying the Balance Law to a three-compartment model to obtain a linear differential system. People seem to enjoy this model, perhaps because it shows the vulnerability of living in Southern California (at least in the bad old days of leaded gasoline and leaded paint). The model provides an excellent backdrop for illustrating the asymptotic approach to equilibrium, the effect of a discontinuous driving term, and the sensitivity of the system as parameters in the model change. Turn ahead to Section 7.10 and the corresponding part of the Resource Manual for more on the lead model and a discussion of general compartment models. Here are two more references on the lead model: *Getting the Lead Out*, Irene Kessel and John T. O'Connor [Plenum, New York, 1997]; "Biological Modeling for Predictive Purposes," Naomi H. Harley, pp. 27–36 in *Methods for Biological Monitoring*, T.J. Kneip and J.V. Crable (editors) [American Public Health Association, Washington, 1988].

---

1. We want to find the equilibrium points of IVP (4):

$$x_1' = -0.0361x_1 + 0.0124x_2 + 0.000035x_3 + 49.3, \qquad x_1(0) = 0$$
$$x_2' = 0.0111x_1 - 0.0286x_2, \qquad x_2(0) = 0$$
$$x_3' = 0.0039x_1 - 0.000035x_3, \qquad x_3(0) = 0$$

(a). Set the rate equations of IVP (4) equal to 0 and solve simultaneously for x_1, x_2, x_3, The equilibrium levels are found to be $x_1 = 1800$, $x_2 = 699$, and $x_3 = 200583$.

(c). Replace 49.3 by I_1 in the above rate equations. The equilibrium levels, $x_1 = 36.5I_1$, $x_2 = 14.16I_1$, and $x_3 = 4068.6I_1$, are obtained by setting the right-hand sides of the ODEs in (4) (with 49.3 replaced by I_1) equal to 0, and solving for x_1, x_2, and x_3. Since we are solving a linear algebraic system of the form $Ax = F$ with solution $x = \hat{x}$, say, then $Ax = F/2$ has solution $x = \hat{x}/2$ by linearity.

3. Begin with IVP (4):

$$x_1' = -0.0361x_1 + 0.0124x_2 + 0.000035x_3 + 49.3, \qquad x_1(0) = 0$$
$$x_2' = 0.0111x_1 - 0.0286x_2, \qquad x_2(0) = 0$$
$$x_3' = 0.0039x_1 - 0.000035x_3, \qquad x_3(0) = 0$$

(a). Replacing 49.3 by $49.3\,\text{step}(400 - t) + 24.65\,\text{step}(t - 400)$ models cutting the intake rate in half because at $t = 400$ days the second step function turns on, and the first one turns off; for $t \geq 400$ the intake rate is 24.65 micrograms/day. Fig. 3(a) shows that by $t = 800$ days, the lead levels in the blood and the tissues have been cut in half.

(c). Fig. 3(c) shows the significant changes in blood and tissues lead levels if the intake rate is cut in half after 400 days [part **(a)**] and medication doubles the clearance coefficient k_{01} [part **(b)**]. The new rate equation for x_1 is $x_1' = -[0.0361 \, \text{step}(400 - t) + 0.0572 \, \text{step}(t - 400)]x_1 + 0.0124x_2 + 0.000035x_3 + 49.3 \, \text{step}(400 - t) + 24.65 \, \text{step}(t - 400)$.

Problem 3(a).

Problem 3(c).

5. **(a).** Here is the compartment diagram for the flow of inulin through the bloodstream (compartment 1) and the intercellular areas (compartment 2):

Suppose that $x_1(t)$, $x_2(t)$ are the respective amounts of inulin at time t in the blood and in the intercellular area. The Balance Law (1) implies that $x_1' = I_1 - (k_{21} + k_{01})x_1 + k_{12}x_2$, $x_2' = k_{21}x_1 - k_{12}x_2$.

(c). See Fig. 5(c), where the respective equilibrium levels of inulin from the formulas of part **(b)** are $x_1 = 200$, $x_2 = 100$. The x_1 curve (solid) and x_2 curve (dashed) are graphed over the time span $0 \leq t \leq 1500$, but each reaches the 90% equilibrium level long before $t = 1500$ (around $t = 750$).

Problem 5(c).

7.2 Overview of Vectors and Matrices

Comments

We avoid the axiomatic definition of a vector space. Instead, we define a vector space as the collection of all column vectors with n entries along with matrix addition and multiplication by a scalar. It is important to distinguish the case where all entries in the vectors (and the scalars) are reals (i.e., the case $\mathbb{R}^n$) from the case where the entries (and the scalars) are complex numbers (i.e., the case $\mathbb{C}^n$). We use the term function vector space to mean the collection of all n-entry column vectors with function entries (the notion $\mathbf{C}^n(I)$ is used again, leaving the number of entries to the context).

It is important to take a look at matrices with complex entries and the corresponding linear systems with complex coefficients. Ditto for function vector spaces with entries which are complex-valued functions of a real variable.

1. Here's the matrix equation we want to solve for A:

$$2A + \begin{bmatrix} 0 & 1 \\ 1 & -1 \end{bmatrix} A = \begin{bmatrix} 1 & -2 \\ 0 & 1 \end{bmatrix}$$

Put $A = \begin{bmatrix} a & b \\ c & d \end{bmatrix}$ into the given equation to obtain

$$\begin{bmatrix} 2a & 2b \\ 2c & 2d \end{bmatrix} + \begin{bmatrix} c & d \\ a-c & b-d \end{bmatrix} = \begin{bmatrix} 1 & -2 \\ 0 & 1 \end{bmatrix}$$

Comparing entries, we conclude that

$$2a + c = 1, \quad a + c = 0, \quad 2b + d = -2, \quad b + d = 1$$

Solving this system, we obtain the unique solution $a = 1$, $b = -3$, $c = -1$, $d = 4$.

3. The proof is by direct computation. The i-th entry in the column vector $x_1 a^1 + \cdots + x_n a^n$ is $\sum_{j=1}^n a_{ij} x_j$, but this is also the i-th entry of the matrix product Ax. We leave the examples to the reader.

5. **(a).** Let v^1 and v^2 be any two vectors common to the subspaces U and W of the space V, and let α, β be any two scalars. Then $\alpha v^1 + \beta v^2$ must belong to U and also to W because they are subspaces so each contains all finite linear combinations of its own elements. Therefore, $\alpha v^1 + \beta v^2$ is common to U and W, andthe vectors common to U and W form a subspace of V (by the definition of subspace).

7. **(a).** The given system is linear and has the standard form

$$\begin{bmatrix} x_1 \\ x_2 \\ x_3 \end{bmatrix}' = \begin{bmatrix} 1 & 1 & -2 \\ -5 & -1 & 2 \\ -1 & -2 & 1 \end{bmatrix} \begin{bmatrix} x_1 \\ x_2 \\ x_3 \end{bmatrix} + \begin{bmatrix} 0 \\ e^{-t} \\ \sin t \end{bmatrix}$$

7.3   Systems of Linear Equations

Comments

There are some subtle ideas here, and it may take some time for everything to sink in. Many readers have already seen determinants and linear systems before in some math course, but they might be a bit rusty on these topics. The main point is how to find *all* solutions of a linear system, even when complex numbers are involved (we will need this later). The other main point is linear independence and how to find a basis for a subspace. Work as many examples as you can and read over the many examples in this section.

1. **(a).** Write the system in standard matrix form $\begin{bmatrix} 1 & -2 \\ -2 & 4 \end{bmatrix} \begin{bmatrix} x_1 \\ x_2 \end{bmatrix} = \begin{bmatrix} 0 \\ 0 \end{bmatrix}$ and use ele-

mentary row operations on the augmented matrix $\left[\begin{array}{cc|c} 1 & -2 & 0 \\ -2 & 4 & 0 \end{array}\right]$ to obtain $\left[\begin{array}{cc|c} 1 & -2 & 0 \\ 0 & 0 & 0 \end{array}\right]$.

So the solution set of our system is given by the single equation $x_1 - 2x_2 = 0$. This equation determines a line through the origin in the $x_1 x_2$-plane. To describe this line assign the arbitrary value s to x_2. Then $x_1 = 2s$, and the solution set is given by

$$x = \begin{bmatrix} x_1 \\ x_2 \end{bmatrix} = \begin{bmatrix} 2s \\ s \end{bmatrix} = s \begin{bmatrix} 2 \\ 1 \end{bmatrix}$$

(c). Writing the system in standard augmented matrix form we have

$$\left[\begin{array}{ccc|c} 1 & 1 & 2 & 1 \\ 3 & 4 & -1 & -1 \end{array}\right]$$

and using elementary row operations we have

$$\left[\begin{array}{ccc|c} 1 & 1 & 2 & 1 \\ 0 & 1 & -7 & -4 \end{array}\right]$$

which yields the two equations $x_1 + x_2 + 2x_3 = 1$, and $x_2 - 7x_3 = -4$. Assigning the arbitrary value s to x_3, the second equation reduces to $x_2 = -4 + 7s$, and the first equation becomes $x_1 = 5 - 9s$. So the solution set of our system is

$$x = \begin{bmatrix} x_1 \\ x_2 \\ x_3 \end{bmatrix} = \begin{bmatrix} 5 - 9s \\ -4 + 7s \\ s \end{bmatrix} = \begin{bmatrix} 5 \\ -4 \\ 0 \end{bmatrix} + s \begin{bmatrix} -9 \\ 7 \\ 1 \end{bmatrix}$$

which is a line in $x_1 x_2 x_3$-space passing through the point $(5, -4, 0)$ and parallel to the vector $-9\mathbf{i} + 7\mathbf{j} + \mathbf{k}$.

3. **(a).** We omit the calculations that carry us to the reduced system from which the solution can be found. Applying elementary row operations to the augmented matrix of the system, we obtain

$$\begin{bmatrix} 3 & -1 & 1 & 2 & -2 & 1 \\ 0 & -7 & 1 & 5 & -5 & 3 \\ 0 & 0 & 15 & -30 & 9 & 17 \end{bmatrix}$$

which translates to the system

$$3x_1 - x_2 + x_3 + 2x_4 - 2x_5 = 1$$

$$-7x_2 + x_3 + 5x_4 - 5x_5 = 3$$

$$15x_3 - 30x_4 + 9x_5 = 17$$

Putting $x_4 = a$, $x_5 = b$ for arbitrary reals a and b, we see that $x_3 = 17/15 + 2a - (3/5)b$, $x_2 = 2/35 + a - (4/5)b$, and $x_1 = -8/315 - a + (3/5)b$. So all solutions of the system are given by

$$x = \begin{bmatrix} x_1 \\ x_2 \\ x_3 \\ x_4 \\ x_5 \end{bmatrix} = \begin{bmatrix} -8/315 - a + (3/5)b \\ 2/35 + a - (4/5)b \\ 17/15 + 2a - (3/5)b \\ a \\ b \end{bmatrix} = \begin{bmatrix} -8/315 \\ 2/35 \\ 17/15 \\ 0 \\ 0 \end{bmatrix} + a \begin{bmatrix} -1 \\ 1 \\ 2 \\ 1 \\ 0 \end{bmatrix} + b \begin{bmatrix} 3/5 \\ -4/5 \\ -3/5 \\ 0 \\ 1 \end{bmatrix}$$

for arbitrary reals a and b.

5. The upper triangular matrix $\begin{bmatrix} a & b \\ 0 & c \end{bmatrix}$ is invertible if and only if there exists a matrix $\begin{bmatrix} A & B \\ C & D \end{bmatrix}$ such that

$$\begin{bmatrix} A & B \\ C & D \end{bmatrix} \begin{bmatrix} a & b \\ 0 & c \end{bmatrix} = \begin{bmatrix} a & b \\ 0 & c \end{bmatrix} \begin{bmatrix} A & B \\ C & D \end{bmatrix} = \begin{bmatrix} 1 & 0 \\ 0 & 1 \end{bmatrix}$$

So, in particular, $Ca = 0$, $Aa = 1$, and $cC = 0$, $cD = 1$. So $a \neq 0$, and $c \neq 0$, which implies that $C = 0$, and we are done.

7. **(a).** If a matrix A is invertible, then there is a matrix M such that $MA = AM = I$. Say that there is another matrix N such that $NA = AN = I$. Then $M = MI = MAN = IN = N$. So inverses are unique.

9. **(a).** The set $\{e^{-t}, 3e^t, \cosh t\}$ is dependent, since $\cosh t = \frac{1}{2}(e^{-t}) - \frac{1}{6}(-3e^t) = \frac{1}{2}(e^t + e^{-t})$.

(c). The set $\{e^{-t}\cos t, e^{-t}\sin t\}$ is independent, for let scalars a, b be such that $ae^{-t}\cos t + be^{-t}\sin t = 0$ for all t. Then it would follow (after dividing by e^{-t}) that $a\cos t + b\sin t = 0$ for all t. Substituting $t = 0$ and $t = \pi/2$ into this statement we obtain that $a = b = 0$, so the set is independent.

11. **(a).** The set of polynomials of degree ≥ 2 is *not* a subspace since it doesn't contain the zero function.

(c). Since any linear combination of odd functions is odd, it follows that the odd functions in $\mathbf{C}^0[-1, 1]$ form a subspace.

13. With the matrix A in Problem 2 we see from Problem 2 **(c)** that $Ax = y$ is solvable if and only if $c - 3a - 2b = 0$. Now let's find $N(A^T)$, that is, the set of all vectors $v = [v_1 \ v_2 \ v_3]^T$ such that $A^T v = 0$. Setting up the augmented matrix

$$[A^T|0] = \begin{bmatrix} 1 & -1 & 1 & | & 0 \\ -1 & 3 & 3 & | & 0 \\ 0 & -2 & -4 & | & 0 \end{bmatrix}$$

and applying elementary row operations, we obtain

$$\begin{bmatrix} 1 & -1 & 1 & | & 0 \\ 0 & 2 & 4 & | & 0 \\ 0 & 0 & 0 & | & 0 \end{bmatrix}$$

So the system $A^T v = 0$ translates into the system

$$v_1 - v_2 + v_3 = 0$$

$$2v_2 + 4v_3 = 0$$

Assigning the arbitrary real value of s to v_3, we see that $v_2 = -2s$, and $v_1 = -3s$. So $A^T v = 0$ if and only if $v = s[-3 \ \ -2 \ \ 1]^T$. Setting $y = [a \ \ b \ \ c]^T$, we see that "$y^T v = 0$ for all v in $N(A^T)$" just means that $-3a - 2b + c = 0$ because $v = [-3 \ \ -2 \ \ 1]^T$ is a basis for $N(A^T)$.

7.4   Eigenvalues and Eigenvectors of Matrices

Comments

Look ahead to Section 7.5 and read the discussion just before Example 7.5.1 about the need to do eigen-analysis in order to generate the solution set of $x' = Ax$, where A is a constant matrix. However, most of the section is devoted to the process of finding eigenvalues and eigenvectors, maybe even using a CAS for larger matrices. Everyone needs to understand about eigenspaces and multiple eigenvalues, and at least have a basic idea about generalized eigenvectors. The results of Problems 5–7 are important in later sections. People sometimes have trouble spanning an eigenspace, probably because of having to deal with $(A - \lambda I)v = 0$, where $A - \lambda I$ is singular. Several examples will usually clear this up. A worked-out example also helps with complex eigenvalues or eigenvectors.

1. The eigenvalues of a matrix A are the roots of the characteristic polynomial $p(\lambda) = \det[A - \lambda I]$. In each solution below we begin with the calculation of $p(\lambda)$, followed by the listing of the eigenvalues (i.e., the roots of $p(\lambda)$ along with their multiplicities). Any nonzero constant multiple of an eigenvector is also an eigenvector, so the eigenvectors given below are not unique. If λ_j is a root of multiplicity m_j, then the corresponding eigenspace has dimension d_j, $1 \le d_j \le m_j$. A basis for the eigenspace consists of d_j independent vectors v such that $(A - \lambda_j I)v = 0$. In each case the spanning vectors are not unique. Details are given in only a few cases.

    (a). $p(\lambda) = \det \begin{bmatrix} 1-\lambda & 0 \\ 2 & 1-\lambda \end{bmatrix} = (1-\lambda)^2$; $\lambda = 1$ is a double eigenvalue. The corresponding eigenspace V_1 may have dimension 1 or 2. In this case the eigenvector equation $(A - I)v = 0$ reduces to the single equation $2v_1 = 0$. V_1 is spanned by the eigenvector $[0\ 1]^T$ and is one-dimensional.

    (c). $p(\lambda) = \det \begin{bmatrix} -\lambda & 3 \\ 3 & -\lambda \end{bmatrix} = \lambda^2 - 9 = (\lambda - 3)(\lambda + 3)$. The eigenvalues 3 and -3 are simple. V_3 is spanned by $[1\ 1]^T$, V_{-3} by $[1\ -1]^T$.

    (e). $p(\lambda) = \det \begin{bmatrix} 3-\lambda & -5 \\ 5 & 3-\lambda \end{bmatrix} = (3-\lambda)^2 + 25 = \lambda^2 - 6\lambda + 34$. The simple eigenvalues are the complex conjugates $3 + 5i$ and $3 - 5i$. V_{3+5i} is spanned by $[1\ -i]^T$ and V_{3-5i} by the conjugate $[1\ i]^T$.

    (g). Expanding $\det(A - \lambda I) = \det \begin{bmatrix} 5-\lambda & -6 & -6 \\ -1 & 4-\lambda & 2 \\ 3 & -6 & -4-\lambda \end{bmatrix}$ by cofactors (see Section 7.2), we obtain $p(\lambda) = -\lambda^3 + 5\lambda^2 - 8\lambda + 4 = -(\lambda - 1)(\lambda - 2)^2$, so 1 is a simple eigenvalue and 2 is a double eigenvalue. V_1 is spanned by $[3\ -1\ 3]^T$ and V_2 has basis $\{[2\ 1\ 0]^T, [2\ 0\ 1]^T\}$. Any two independent vectors in V_2 form a basis, so there is nothing unique about the given basis.

    (i). $p(\lambda) = \det \begin{bmatrix} -1-\lambda & 36 & 100 \\ 0 & -1-\lambda & 24 \\ 0 & 0 & 5-\lambda \end{bmatrix} = (-1-\lambda)^2(5-\lambda)$; -1 is a double eigenvalue and 5 is a simple eigenvalue. V_{-1} is spanned by $[1\ 0\ 0]^T$ and V_5 is spanned by $[262\ 27\ 6]^T$ since the eigenvector equation $(A - 5I)v = 0$ reduces to $-6v_1 + 36v_2 + 100v_3 = 0$, $-6v_2 + 27v_3 = 0$.

3. If $Av = \lambda v$ for some vector $v \ne 0$, then $A^k v = A^{k-1}(Av) = A^{k-1}(\lambda v) = \lambda A^{k-1}(v) = \cdots = \lambda^k v$. So λ^k is an eigenvalue of A^k if λ is an eigenvalue of A. The other way around, however, is not always true. For example, $A = \begin{bmatrix} 2 & 0 \\ 0 & -3 \end{bmatrix}$ has eigenvalues 2 and -3, while $A^2 = \begin{bmatrix} 4 & 0 \\ 0 & 9 \end{bmatrix}$ has eigenvalues 4 and 9, but $3 = 9^{1/2}$ is *not* an eigenvalue of A.

5. We have $\det[A - \lambda I] = p(\lambda) = (-1)^n(\lambda - \lambda_1) \cdots (\lambda - \lambda_n)$ if $\lambda_1, \ldots, \lambda_n$ are the eigenvalues of A, each repeated according to its multiplicity (see "Polynomials" in Appendix B.4). So $p(0) = (-1)^n(-\lambda_1) \cdots (-\lambda_n) = \lambda_1 \cdots \lambda_n = \det(A - 0I) = \det A$. On the other hand, the constant term of $p(\lambda) = \det(A - \lambda I)$ is $p(0) = \det A$, so $\det A$ is the product of its eigenvalues.

7. By Problem 5, $\det A = \lambda_1 \cdots \lambda_n$, where $\lambda_1, \ldots, \lambda_n$ are the eigenvalues of A. Since A is nonsingular if and only if its determinant is nonzero (see Theorem 7.3.1), we see that A is nonsingular if and only if none of its eigenvalues are zero.

9. First we find the eigenvalues of the given matrix:

$$p(\lambda) = \det \begin{bmatrix} -2 - \lambda & -2 & 0 \\ 2 & 3 - \lambda & 0 \\ 0 & 0 & -1 - \lambda \end{bmatrix} = (\lambda^2 - \lambda - 2)(-1 - \lambda) = -(\lambda - 2)(\lambda + 1)^2;$$

$\lambda = 2$ is a simple eigenvalue and $[1 \quad -2 \quad 0]^T$ is a corresponding eigenvector; $\lambda = -1$ is a double eigenvalue, and V_{-1} is 2-dimensional and is spanned by $[2 \quad -1 \quad 0]^T$ and $[2 \quad -1 \quad 1]^T$. So the basis of $\mathbb{R}^3$ consists of three eigenvectors, and we don't need any generalized eigenvectors.

7.5 Undriven Linear Systems with Constant Coefficients

Comments

With the matrix and eigen machinery of Sections 7.2–7.4 in hand, we can solve undriven, constant coefficient linear systems, and that is what we do in this section. Operator notation is used first to simplify matters, and to help explain why the process of finding all solutions of $x' = A(t)x + F(t)$ splits two subproblems: find all solutions of $L[x] = 0$, where L is the linear operator $L = d/dt - A$ [i.e., first find the general solution of $x' = A(t)x$], and then find a particular solution of $Lx = f$. In this section we show how the first problem can be solved if A is a matrix of constants which has all real eigenvalues and the eigenvalues of A (and their multiplicities) are known, as well as the corresponding eigenspaces. Wronskians and fundamental sets are introduced. The only real algebraic problem in all this occurs when the dimension of the eigenspace corresponding to a multiple eigenvalue is less than the multiplicity of the eigenvalue. The generalized eigenvectors introduced in Section 7.4 come to the rescue in this case. We work out a few examples using a chain of generalized eigenvectors. We don't dwell on this awkward case, but focus on the more common situation of full eigenspaces (we call them nondeficient in the text).

1. In each case we write out the system matrix rather than the full system of ODEs. That's all that is needed to construct the solutions (using Theorem 7.5.3).

(a). The system matrix is $A = \begin{bmatrix} 5 & 3 \\ -1 & 1 \end{bmatrix}$. Since $\det(A - \lambda I) = p(\lambda) = \lambda^2 - 6\lambda + 8 = (\lambda - 2)(\lambda - 4)$, the eigenvalues are $\lambda_1 = 2$ and $\lambda_2 = 4$. To find the eigenspace for λ_1, note that $(A - 2I)v = 0$ reduces to $3v_1 + 3v_2 = 0$, so V_2 is spanned by $[1 \quad -1]^T$. For λ_2, note that $(A - 4I)v = 0$ reduces to $v_1 + 3v_2 = 0$ so V_4 is spanned by $[3 \quad -1]^T$. The general real solution of the given system is

$$x = c_1 \begin{bmatrix} 1 \\ -1 \end{bmatrix} e^{2t} + c_2 \begin{bmatrix} 3 \\ -1 \end{bmatrix} e^{4t} \qquad c_1, c_2 \text{ arbitrary reals}$$

(c). The system matrix is $A = \begin{bmatrix} 1 & -1 \\ 1 & 3 \end{bmatrix}$ and since $\det(A - \lambda I) = p(\lambda) = (\lambda - 2)^2$, $\lambda_1 = 2$ is the only eigenvalue and it has multiplicity $m_1 = 2$. To find the eigenspace for λ_1, note that $(A - 2I)v = 0$ becomes $v_1 + v_2 = 0$. So the eigenspace for λ_1 is spanned by $[1 \quad -1]^T$. So $x^1 = e^{2t}[1 \quad -1]^T$ is one solution of the given system. Since the eigenspace is deficient, we need to find a generalized eigenvector in order to construct a fundamental set. Any vector w which solves $(A - 2I)w = [1 \quad -1]^T$ is a generalized eigenvector. Solving, we have $w = [0 \quad -1]^T$ as one such solution. Now using the notation of (16) for a chain of eigenvectors, we put $u^1 = [1 \quad -1]^T$ and $u^2 = [0 \quad -1]^T$, so from (18), $x^2 = ([0 \quad -1]^T + t[1 \quad -1]^T)e^{2t}$ is another solution of the given system. Also, $\{x^1, x^2\}$ is a fundamental set, so $x = c_1 x^1 + c_2 x^2$, c_1 and c_2 arbitrary reals, is the general solution of the given system.

3. **(a).** The eigenvalues of the system matrix $\begin{bmatrix} 0 & 1 \\ 8 & -2 \end{bmatrix}$ are $\lambda_1 = -4$, $\lambda_2 = 2$. The eigenspace for λ_1 is spanned by $[1 \quad -4]^T$, and the eigenspace for λ_2 is spanned by $[1 \quad 2]^T$. The general real-valued solution of the system is $x = c_1 e^{-4t}[1 \quad -4]^T + c_2 e^{2t}[1 \quad 2]^T$, where c_1 and c_2 are arbitrary reals.

(c). The eigenvalues of the system matrix $\begin{bmatrix} 2 & 1 \\ -3 & 6 \end{bmatrix}$ are $\lambda_1 = 5$ and $\lambda_2 = 3$, and the corresponding eigenspaces are spanned, respectively, by $[1 \quad 3]^T$ and $[1 \quad 1]^T$. So the general real-valued solution is $x = c_1 e^{5t}[1 \quad 3]^T + c_2 e^{3t}[1 \quad 1]^T$, where c_1 and c_2 are arbitrary reals.

(e). The eigenvalues of the system matrix $\begin{bmatrix} 3 & 2 & 2 \\ 1 & 4 & 1 \\ -2 & -4 & -1 \end{bmatrix}$ are $\lambda_1 = 1$, $\lambda_2 = 2$, $\lambda_3 = 3$, and the corresponding eigenspaces are spanned by $[1 \quad 0 \quad -1]^T$, $[2 \quad -1 \quad 0]^T$ and $[0 \quad 1 \quad -1]^T$, respectively. So the general real-valued solution is

$$x = c_1 e^t[1 \quad 0 \quad -1]^T + c_2 e^{2t}[2 \quad -1 \quad 0]^T + c_3 e^{3t}[0 \quad 1 \quad -1]^T$$

where c_1, c_2, and c_3 are arbitrary reals.

5. **(a).** Note that $x_i = y^{(i-1)}$, and so $x_i' = y^{(i)} = x_{i+1}$, for $i = 1, 2, \ldots, n - 1$. Finally, using the ODE,

$$x_n' = y^{(n)} = -a_0 y - a_1 y' - \cdots - a_{n-1} y^{(n-1)}$$

$$= -a_0 x_1 - a_1 x_2 - \cdots - a_{n-1} x_n$$

The vector $[x_1' \quad x_2' \cdots x_n']^T$ takes the form $A[x_1 \quad x_2 \cdots x_n]^T$, where the matrix has the asserted form.

7. Group project, but here is a sketch of how to do it.

- Because A is a triangular matrix, we immediately have

$$\det(A - \lambda I) = (-1 - \lambda)^3 = -(\lambda + 1)^3$$

and so A has only one eigenvalue $\lambda_1 = -1$; its multiplicity is $m_1 = 3$. To find the eigenspace we look for all solutions of the algebraic system $(A + I)v = 0$. All solutions of this equation have the form

$$v = \begin{bmatrix} a \\ b \\ 0 \end{bmatrix} = a \begin{bmatrix} 1 \\ 0 \\ 0 \end{bmatrix} + b \begin{bmatrix} 0 \\ 1 \\ 0 \end{bmatrix}$$

for arbitrary reals a and b. The eigenspace V_{-1} has dimension two because it is spanned by the basis $\{[1 \ \ 0 \ \ 0]^T, [0 \ \ 1 \ \ 0]^T\}$.

- Let the eigenvector u^1 have the general form $u^1 = [c_1, \ c_2, \ 0]^T$. Find constants c_1, c_2 such that the equation $(A + I)u^2 = [c_1 \ \ c_2 \ \ 0]^T$ has a solution for u^2. Writing this equation in component form (with $u^2 = [u_1^2 \ \ u_2^2 \ \ u_3^2]^T$), we have that $2u_3^2 = c_1$ and $u_3^2 = c_2$, which is solvable for any pair c_1, c_2, with $2c_2 = c_1$. So taking $c_1 = 2$, $c_2 = 1$, and solving we get $u_3^2 = 1$. There are no restrictions on u_1^2 and u_2^2, and so take them both to be zero. We have the chain

$$u^1 = [2 \ \ 1 \ \ 0]^T, \quad \text{and} \quad u^2 = [0 \ \ 0 \ \ 1]^T$$

- We have three solutions of $x' = Ax$,

$$x^1 = e^{-t}[1 \ \ 0 \ \ 0]^T$$
$$x^2 = e^{-t}[2 \ \ 1 \ \ 0]^T$$
$$x^3 = e^{-t}([0 \ \ 0 \ \ 1]^T + t[2 \ \ 1 \ \ 0]^T)$$

The Wronskian $W[x^1, x^2, x^3]$ evaluated at $t = 0$ has the value $W(0) = 1$, and so $\{x^1, x^2, x^3\}$ is a basic solution set. So the general real-valued solution of $x' = Ax$ is $x = c_1 x^1 + c_2 x^2 + c_3 x^3$, where c_1, c_2, c_3 are arbitrary reals.

7.6 Undriven Linear Systems: Complex Eigenvalues

Comments

The reason why we put this material in a separate section is because people are often uncomfortable working with matrices and vectors whose entries are complex numbers or complex-valued functions. The matter-of-fact approach in Section 7.5 also works in the complex eigenvector case if the algebra of complex numbers is used. Our "proof" of solution formula (3) is by construction, once a basis of eigenvectors and generalized eigenvectors is known. We did not prove the existence of such a basis for the general case (because that would be beyond the scope of this text—or even a first course in Linear Algebra). This is not a problem because our constructive approach works just fine when $n \leq 4$, and these are the examples that occur in this text. Basic solution sets are important, so is Theorem 7.6.4 because it tells us the form of every term in every component of every solution vector of $x' = Ax$.

1. The eigenvalues of the system matrix $A = \begin{bmatrix} 2 & -1 \\ 1 & 2 \end{bmatrix}$ are $2 \pm i$. The corresponding eigenspaces are spanned by $[1 \ \mp i]^T$. The general solution of $x' = Ax$ is

    $$x = c_1[1 \ -i]^T e^{(2+i)t} + c_2[1 \ i]^T e^{(2-i)t},$$

    where c_1, c_2 are arbitrary complex numbers. To find the general real-valued solutions, put $x^1 = \text{Re}\left[[1 \ -i]^T e^{(2+i)t}\right] = e^{2t}[\cos t \ \sin t]^T$, and $x^2 = \text{Im}\left[[1 \ -i]^T e^{(2+i)t}\right] = e^{2t}[\sin t \ -\cos t]^T$. Then $x = c_1 x^1 + c_2 x^2$, c_1 and c_2 arbitrary reals, is the general real solution.

3. **(a).** The eigenvalues of the system matrix $A = \begin{bmatrix} 0 & 1 \\ -64 & 0 \end{bmatrix}$ are $\lambda_1 = 8i$, $\lambda_2 = -8i$. The eigenspaces are spanned, respectively, by $[1 \ 8i]^T$ and $[1 \ -8i]^T$. So the general (complex-valued) solution of $x' = Ax$ is $x = c_1 e^{8it}[1 \ 8i]^T + c_2 e^{-8it}[1 \ -8i]^T$. To find all real-valued solutions, put

    $$x^1 = \text{Re}\left[e^{8it}[1 \ 8i]^T\right] = [\cos 8t \ -8\sin 8t]^T$$

    $$x^2 = \text{Im}\left[e^{8it}[1 \ 8i]^T\right] = [\sin 8t \ 8\cos 8t]^T$$

    Then the general real-valued solution is $x = c_1 x^1 + c_2 x^2$, where c_1 and c_2 are arbitrary reals.

5. **(a).** Suppose that A is a real matrix with a complex eigenvalue λ. Now since $Av = \lambda v$, we see by taking conjugates that $\overline{Av} = \overline{\lambda} \overline{v}$. But since A is real we have that $\overline{Av} = A\overline{v}$, so $A\overline{v} = \overline{\lambda} \overline{v}$. So $\overline{v}$ is an eigenvector of A corresponding to the eigenvalue $\overline{\lambda}$. Now if v had only real components then $v = \overline{v}$, so

    $$\lambda v = Av = A\overline{v} = \overline{\lambda} \overline{v} = \overline{\lambda} v$$

    We see that $\lambda = \overline{\lambda}$, a contradiction since λ is assumed to be nonreal. So v must have at least one nonreal component.

    (c). Let $z = ve^{\lambda t}$, where v and λ are as in part **(a)**. Then z and $\overline{z} = \overline{v}e^{\overline{\lambda} t}$ are complex-valued solutions of $x' = Ax$, and from part **(b)**, $\{z, \overline{z}\}$ is an independent set. Setting $x^1 = \text{Re}[z]$ and $x^2 = \text{Im}[z]$, observe that

    $$x^1 = \frac{1}{2}(z + \overline{z}) \text{ and } x^2 = \frac{-i}{2}(z - \overline{z})$$

    so $\{x^1, x^2\}$ are real-valued solutions of $x' = Ax$. Now let the constants a and b be such that $ax^1 + bx^2 = 0$. We show then that $a = b = 0$, and $\{x^1, x^2\}$ is an independent set. Indeed, replacing x^1 and x^2 by their equivalents in terms of z, $\overline{z}$, we see that $(a+b)z + (a-b)\overline{z} = 0$. But since $\{z, \overline{z}\}$ are independent we have that $a - bi = a + bi = 0$, so $a = b = 0$, finishing the proof.

7.7   Orbital Portraits

Comments

The focus of this section is on the gallery of the portraits of the orbits of linear planar autonomous systems. There are six qualitatively different portraits here (more if we include system matrices with one or more zero eigenvalues). The saddle, spiral point, center, and three kinds of nodal portraits are by now visually distinct icons for the systems they portray. Then we show how a planar orbital portrait evolves as one of the coefficients in the rate function changes. Finally, we consider a very different question, the *second-order* system $z'' = Az$, where A is an $n \times n$ matrix of real constants, and what its solution set must look like. This leads us to Theorem 7.7.1, which characterizes the general solution $z(t)$ of $z'' = Az$ in terms of the eigenvectors of A and the *square roots* of the eigenvalues in the case that the eigenvalues of A are nonzero and distinct. Rather than reduce $z'' = Az$ to a system of $2n$ first order linear ODEs by introducing the n velocities $z'_1, \ldots, z'_n$ as additional state variables, we treat the second-order system directly. That way we can talk about normal modes and frequencies in the way physicists, physical chemists, and mechanical engineers do. The model of the coupled springs is illustrative. It is important to realize that orbits really live in the $2n$-dimensional state space of z and z'. The normal mode oscillations and Lissajous curves that seem to cross one another in the n-dimensional z space are actually the projections of orbits in zz'-space down into z-space.

The emphasis throughout the section is on relating the orbital portraits to the nature of the eigensets of the system matrix. Incidentally, here is a good spot to focus on the art of creating good graphics—choosing the right ranges for the variables, selecting good initial points, drawing enough curves to be informative but not so many that the picture is cluttered. We have tried to follow these priniciples in our pictures. This is also a good place to look at an orbital portrait and ask yourself what the portrait implies about the eigenvalues (and eigenvectors) of the unknown system. This makes a good exam problem!

---

1. From Example 7.7.7 we know that the characteristic polynomial of the system matrix $\begin{bmatrix} \alpha & 1 \\ -1 & -1 \end{bmatrix}$ is $p(\lambda) = \lambda^2 + (1 - \alpha)\lambda + 1 - \alpha$, and the eigenvalues are

$$\lambda_{1,2} = \frac{-(1 - \alpha) \pm \sqrt{(1 - \alpha)^2 - 4(1 - \alpha)}}{2} = \frac{-(1 - \alpha) \pm \sqrt{(1 - \alpha)(-\alpha - 3)}}{2}$$

(a). If $\alpha > 1$, we know $\lambda_1 = [-(1 - \alpha) + \sqrt{(1 - \alpha)(-\alpha - 3)}]/2 > 0$ and $\lambda_2 = [-(1 - \alpha) - \sqrt{(1 - \alpha)(-\alpha - 3)}]/2 < 0$, so the equilibrium point at the origin is a saddle because the eigenvalues have opposite signs.

(c). If $\alpha < -3$, $\lambda_{1,2} = -\frac{1}{2}(1 - \alpha) \pm \frac{1}{2}(1 - \alpha)\sqrt{1 - 4(1 - \alpha)^{-1}} < 0$, and $\lambda_1 \neq \lambda_2$. So the origin is an improper node if $\alpha < -3$. Since the eigenvalues are real and distinct, a pair of eigenvectors (one for each eigenvalue) is independent.

3. See the figures for orbital portraits of the systems of Problem 2.

(a). The system is $x' = x + y$, $y' = 4x - 2y$ [Fig. 3(a)].

(c). The system is $x' = 7x + 6y$, $y' = 2x + 6y$ [Fig. 3(c)].

(e). The system is $x' = -x - 4y$, $y' = x - y$ [Fig. 3(e)].

Problem 3(a).

Problem 3(c).

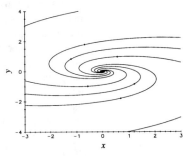

Problem 3(e).

5. The affine system has the form $x' = Ax + [r \ s]^T$, where $x = [x \ y]^T$.

(a). We wish to find (x_0, y_0) such that $x' = 0$ and $y' = 0$. That is, we want to solve the system $Ax = -[r \ s]^T$:

$$A \begin{bmatrix} x_0 \\ y_0 \end{bmatrix} = \begin{bmatrix} a & b \\ c & d \end{bmatrix} \begin{bmatrix} x_0 \\ y_0 \end{bmatrix} = \begin{bmatrix} -r \\ -s \end{bmatrix}$$

Since $\det \begin{bmatrix} a & b \\ c & d \end{bmatrix} \neq 0$ the system has a unique solution,

$$[x_0 \quad y_0]^T = \begin{bmatrix} a & b \\ c & d \end{bmatrix}^{-1} \begin{bmatrix} -r \\ -s \end{bmatrix}$$

(c). This follows directly from **(b)**.

7. **(a).** Using the appropriate trigonometric identities from Appendix B.4, we see that $x_1 = -2 \sin 2t$ and $x_2 = -\cos 2t$. Since $x_1' = 4x_2$ and $x_2' = -x_1$ for this particular solution, we already have the desired system: $x_1' = 4x_2$, $x_2' = -x_1$. Since the eigenvalues of the system matrix are $2i$, $-2i$, we can calculate the eigenvectors, which are, respectively, $[2 \ i]^T$ and $[2 \ -i]^T$. The general real-valued solution is

$$\begin{bmatrix} x_1 \\ x_2 \end{bmatrix} = c_1 \begin{bmatrix} 2 \cos 2t \\ -\sin 2t \end{bmatrix} + c_2 \begin{bmatrix} 2 \sin 2t \\ \cos 2t \end{bmatrix}$$

for arbitrary reals c_1, c_2. The solution originally given corresponds to $c_1 = 0$, $c_2 = -1$.

9. Group project.

7.8 Driven Systems and the Matrix Exponential

Comments

In this section we characterize the solution set of the linear system $x' = Ax + F(t)$, where A is a matrix of constants, using a variation of parameters approach. Basic matrices and transition matrices make their appearance—all quite abstract, but leading to the elegant formula (16) for the general solution and the even nicer formula (17) for the unique solution of the corresponding IVP. To find all solutions of the undriven system, $x' = Ax$, one can use any basic solution matrix $X(t)$, and write the general solution as $x(t) = X(t)c$, where c is any constant vector. On the other hand, in order to characterize the solution of the driven IVP as the sum of the responses to the initial data and the driving force as we do in formula (17), you need to use the transition matrix. Given any basic solution matrix $X(t)$, one can construct the transition matrix as $X(t)X^{-1}(t_0)$ [see formula (8)].

Finally, we get to the other topic in the title of the section, the matrix exponential. People are often mystified by the matrix exponential. Our approach is to emphasize that e^{tA} is just a name for the basic solution matrix $X(x)$ for which $X(0) = I$. Or write it as $\exp(tA)$, which seems less mysterious. Most of the properties of e^{tA} follow naturally. However, some properties depend on the series formula for e^{tA}. We don't usually prove the convergence of this series, but we do illustrate its use in calculating e^{tA} if A is diagonal or nilpotent. Incidentally, it has to be emphasized that in general the matrix exponential is *not* the matrix of the exponentials of the individual entries. In addition $\exp\left[\int_{t_0}^{t} A(s)\,ds\right]$ is *not* a solution matrix of $x' = A(t)x$ if $A(t)$ is nonconstant. We don't mention this fact in the text, but an example might be worked out. You might use the system $x' = 0$, $y' = 2tx + y$. Once you get the idea of a matrix exponential, you might try out the idea of a matrix logarithm, or a matrix sine, e.g., $\sin A = A - \frac{1}{3!}A^3 + \frac{1}{5!}A^5 - \frac{1}{7!}A^7 + \cdots$, etc., defined by means of series. Although we haven't done so, the study of matrix functions like these would make an interesting group project.

1. First find the eigenvalues λ_1, λ_2 and eigenvectors for the eigenspaces of the system matrix, and generalized eigenvectors if necessary. A basic solution matrix $X(t)$ can then be constructed and $e^{tA} = X(t)X^{-1}(0)$. In every case the solution of $x' = Ax$, $x(0) = [1 \quad 2]^T$ is $x = e^{tA}[1 \quad 2]^T$.

(a). The eigenvalues of $A = \begin{bmatrix} 1 & 3 \\ 1 & -1 \end{bmatrix}$ are $\lambda_1 = -2$, $\lambda = 2$; the eigenspaces are $V_{-2} = \text{Span}[1 \ -1]^T$, $V_2 = \text{Span}[3 \ 1]^T$. So

$$X(t) = \begin{bmatrix} e^{-2t} & 3e^{2t} \\ -e^{-2t} & e^{2t} \end{bmatrix}, \quad X^{-1}(0) = \frac{1}{4}\begin{bmatrix} 1 & -3 \\ 1 & 1 \end{bmatrix}$$

$$e^{tA} = X(t)X^{-1}(0) = \frac{1}{4}\begin{bmatrix} e^{-2t} + 3e^{2t} & -3e^{-2t} + 3e^{2t} \\ -e^{-2t} + e^{2t} & 3e^{-2t} + e^{2t} \end{bmatrix}$$

$$x(t) = \frac{1}{4}[-5e^{-2t} + 9e^{2t} \quad 5e^{-2t} + 3e^{2t}]^T$$

(c). The eigenvalue of $A = \begin{bmatrix} -1 & 1 \\ 0 & -1 \end{bmatrix}$ is -1 and it is a double eigenvalue, but V_{-1} is only one-dimensional. Since the system is upper triangular it is simplest to solve it directly by starting with $x'_2 = -x_2$, inserting its solution $c_2 e^{-t}$ into the ODE, $x'_1 = -x_1 + x_2$, and solving that ODE by integrating factors. We have $x_2 = c_2 e^{-t}$, $x_1 = (c_1 + c_2 t)e^{-t}$. A fundamental matrix is

$$X(t) = \begin{bmatrix} e^{-t} & te^{-t} \\ 0 & e^{-t} \end{bmatrix}$$

obtained by first setting $c_1 = 1$, $c_2 = 0$ then $c_1 = 0$, $c_2 = 1$. Since $X(0) = I = X^{-1}(0)$, we have $X(t) = e^{tA}$, and $x(t) = e^{-t}[1 + 2t \quad 2]^T$.

3. **(a).** Since A is the block matrix $\begin{bmatrix} B & 0 \\ 0 & C \end{bmatrix}$, we can write $x = [y \; z]^T$, where y and z have the appropriate number of components determined by the sizes of the blocks B and C. Then $x' = Ax$ is equivalent to the decoupled pair of systems $y' = By$ and $z' = Cz$, whose respective transition matrices are e^{tB} and e^{tC}: $\begin{bmatrix} e^{tB} & 0 \\ 0 & e^{tC} \end{bmatrix}$ is a basic solution matrix and is, in fact, e^{tA} since it reduces to the identity matrix at $t = 0$.

5. First, find the eigenvalues and spanning sets for the eigenspaces of the system matrix; using these, a basic solution matrix $X(t)$ can be constructed. A particular solution $x^d(t)$ is found as explained in the hint. The general solution of $x' = Ax + F(t)$ is then $x = X(t)c + x^d(t)$, where c is any constant column vector. Here, we calculate $X(t)$ and $x^d(t)$ only.

(a). The eigenvalues of $A = \begin{bmatrix} 2 & -1 \\ 5 & -2 \end{bmatrix}$ are $\lambda_1 = i = \bar{\lambda}_2$; the eigenspaces are $V_i = \text{Span}\{[1 \quad 2 - i]^T\} = \bar{V}_{-i}$. So,

$$X(t) = \begin{bmatrix} e^{it} & e^{-it} \\ (2-i)e^{it} & (2+i)e^{-it} \end{bmatrix}$$

To find a particular solution $[x_1(t) \quad x_2(t)]^T$, let $x_1 = ae^t + b$, $x_2 = ce^t + d$. Then we must have that $x'_1 = ae^t = 2(ae^t + b) - (ce^t + d) + e^t$, $x'_2 = ce^t = 5(ae^t + b) - 2(ce^t + d) + 1$. Matching coefficients of corresponding terms, we have that $a = 2a - c + 1$, $c = 5a - 2c$, and $0 = 2b - d$, $0 = 5b - 2d + 1$. Solving, we have $a = 3/2$, $c = 5/2$, $b = -1$, and $d = -2$. So $x_1 = 3e^t/2 - 1$, $x_2 = 5e^t/2 - 2$ gives the components of a particular solution $x^d(t)$. The general solution is $x = X(t)c + x^d(t)$, where c is any vector $[c_1 \; c_2]^T$.

7. **(a).** Since $L[x'] = sL[x] - x^0$, we may apply the Laplace transform operator L to every term of $x' = Ax + F$ and obtain $sL[x](s) - x^0 = AL[x](s) + L[F](s)$, where we have used the fact that $LA = AL$, so $[sI - A]L[x](s) = x^0 + L[F](s)$.

(c). From **(a)** and **(b)**, we have for all s large enough,

$$L[x](s) = [sI - A]^{-1}x^0 + [sI - A]^{-1}L[F](s)$$

Applying L^{-1}, the inverse Laplace transform operator,

$$x(t) = L^{-1}[(sI - A)^{-1}x^0] + L^{-1}[(sI - A)^{-1}L[F]](s)$$

On the other hand, by Theorem 7.8.4, we see that $x = e^{tA}x^0 + \int_0^t e^{(t-s)A}F(s)\,ds = e^{tA}x^0 +$

$e^{tA} * F(t)$. Upon comparison of the two formulas for $x(t)$, we see that $\mathcal{L}^{-1}[(sI - A)^{-1}] = e^{tA}$ and $e^{tA} * F(t) = \mathcal{L}^{-1}[(sI - A)^{-1}\mathcal{L}[F]]$, so we have still another way to find e^{tA}.

7.9   Steady States of Driven Linear Systems

Comments

This section has a distinctly applications-oriented slant. We focus on linear systems where all solutions tend to a constant or to a periodic steady-state and on conditions that assure us that this will happen. So we mostly require that the input vector F for the system $x' = Ax + F$ be either constant or periodic, and that all eigenvalues of the constant system matrix A have negative real parts. Under these conditions, we get a constant or a periodic steady-state that attracts all solutions as $t \to +\infty$. We go even further and state a bounded input-bounded output theorem (BIBO) that assures that if the system is driven by a bounded input $F(t)$, then the output can never be unbounded. This is a crude kind of stability result and is often termed *engineering stability*.

The second half of the section takes up the question of how we can tell whether the eigenvalues of a matrix have negative real parts without actually determining the eigenvalues. Our aim here is to find simple tests that you can carry out with pencil and paper. The first two tests are used to decide whether the roots of a monic polynomial (i.e., a polynomial with leading coefficient 1 such as $r^4 + 3r^3 + 2r^2 + r + 5$) have only negative real parts. The prime test here is the Routh Test, which is widely used by engineers. The general form of the Routh Test is given in the Background Material; the text version covers only quadratic, cubic, and quartic polynomials. For a proof, see Chapter XV, vol. 2, of *Matrix Theory*, F. R. Gantmacher, Chelsea Publ. Co., 1960. The polynomial we have in mind is the characteristic polynomial of a matrix (after the polynomial has been prepared by dividing by its leading coefficient). But who wants to find the characteristic polynomial of a big matrix? So the second set of tests works directly with the system matrix A. Aside from the trace test which is only a necessary (but not sufficient) condition that all eigenvalues have negative real parts, the Gerschgorin Disks Theorem is the main result here. This theorem is easy to use, and you can draw circles in the complex plane that tell you (more or less) just where the eigenvalues lie. We give the proof in the Background Material.

As we mention briefly in this section, there is an ambiguity in the definition of steady-state. We have taken the hard-line and defined steady-state only when the input vector is a constant or else is periodic. On the other hand, if all eigenvalues of A have negative real parts (as we require), and if $F(t)$ is any bounded column vector for $t \geq 0$, then all solutions of $x' = Ax + F(t)$ tend to one another as $t \to +\infty$. So if $x^1(t)$ and $x^2(t)$ are any two solutions, then $\|x^1(t) - x^2(t)\| \to 0$ as $t \to +\infty$, which suggests that all solutions are steady-state. This is just too inclusive for us, and that is why we have taken the hard-line. Some people define a steady-state as any bounded solution of $x' = Ax + F$, whether or not all other solutions approach it. This also is too inclusive, and certainly is contrary to common-sense applications, where steady-state means a regular and predictable state approached by all other solutions.

1.   According to the rather narrow definition of steady-state that we use, and using Theorem 7.9.1, we only need to check that the eigenvalues of A have negative real parts. If that is the case, there is a steady-state; if not, then there is no steady-state. Observe that $x = 0$ is a constant solution for all systems considered.

(a). The trace of A is $-10 + 4 + 7 = 1$, which is positive; according to Theorem 7.9.8, at least one eigenvalue has positive real part. So, $x = 0$ is a constant solution, but it is not a steady-state.

(c). Since the matrix is triangular, its eigenvalues are the diagonal entries $-10, -1, -1$ (see Problem 2, Section 7.4). So, $x = 0$ is the unique steady-state by the Constant Steady State Theorem (Theorem 7.9.1).

3. According to the Routh Test (Theorem 7.9.7) for the characteristic polynomial $r^2 + ar + b$ of the undriven system $x' = y$, $y' = -bx - ay$, every root has negative real part if and only if $a > 0$ and $b > 0$. The eigenvalues of the system matrix for the equivalent system, $x' = y$, $y' = -bx - ay + f(t)$ have negative real parts. The input for the system is $[0 \quad f(t)]^T$, which is bounded since $||[0 \quad f(t)]^T|| = |f(t)|$ and $f(t)$ is assumed to be bounded. The Bounded Input-Bounded Output Theorem (Theorem 7.9.5) applies, and all solutions of the system are bounded for $t \geq 0$.

5. The general solution of the system $x' = -x + e^{3t}y$, $y' = -y$ is $y = c_2 e^{-t}$, $x = c_1 e^{-t} + (c_2/3)e^{2t}$, and if $c_2 \neq 0$, $|x(t)| \to \infty$ as $t \to +\infty$. The eigenvalues of the system matrix $\begin{bmatrix} -1 & e^{3t} \\ 0 & -1 \end{bmatrix}$ are $-1, -1$, but the Bounded Input-Bounded Output Theorem (Theorem 7.9.5) does not apply since the system matrix is *not* a matrix of constants.

7. All of the polynomials [except for that in part **(c)**] have positive coefficients. So we only need to check the other Routh coefficient conditions (Theorem 7.9.7).

(a). The Routh Test applied to $z^3 + z^2 + 2z + 1$ is that $1 \cdot 2 > 1$, which holds. So all roots have negative real parts.

(c). Since the coefficient of z^3 (and of z^2) is 0, the Routh Test implies that at least one of the roots has a nonnegative real part.

(e). First reduce to the monic polynomial $z^3 + z^2 + (11/12)z + 1/2$ by dividing by 12. The Routh Test requires that $1 \cdot (11/12) > 1/2$, and so all roots have negative real parts.

9. In each figure the upper graphs show the x components of solutions, the lower graphs show the y components.

(a). The system is $x' = y + 5$, $y' = -2x - 3y + 10$. The eigenvalues of the system matrix $\begin{bmatrix} 0 & 1 \\ -2 & -3 \end{bmatrix}$ are -1 and -2. By Theorem 7.9.1, the unique steady-state solution is

$$\begin{bmatrix} x \\ y \end{bmatrix} = -A^{-1}F_0 = -\begin{bmatrix} 0 & 1 \\ -2 & -3 \end{bmatrix}^{-1} \begin{bmatrix} 5 \\ 10 \end{bmatrix} = -\begin{bmatrix} -3/2 & -1/2 \\ 1 & 0 \end{bmatrix}\begin{bmatrix} 5 \\ 10 \end{bmatrix} = \begin{bmatrix} 25/2 \\ -5 \end{bmatrix}$$

See Fig. 9(a) for graphs of the component curves of solutions as they approach the steady state.

(c). The system is $x' = -x + y + \sin t$, $y' = x - y$. The eigenvalues of the system matrix are $-1 \pm i$. We can find the unique periodic steady-state solution (i.e., periodic forced oscillation) by Variation of Parameters or by Undetermined Coefficients. The latter process is easier here. Assume a particular solution of the form $x = A \cos t + B \sin t$,

$y = C\cos t + D\sin t$. Inserting these expressions into the ODEs and matching coefficients of corresponding terms, we have that

$$-A = -B + D + 1, \quad B = -A + C, \quad -C = -B - D, \quad D = -A - C$$

Solving, we have $A = -1/5$, $B = 3/5$, $C = 2/5$, $D = -1/5$. All solutions tend to the steady-state

$$x = -(1/5)\cos t + (3/5)\sin t, \quad y = (2/5)\cos t - (1/5)\sin t$$

See Fig. 9(c).

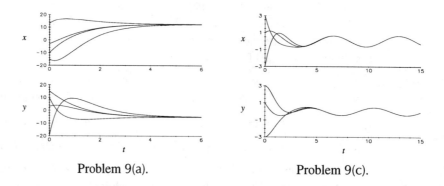

Problem 9(a).   Problem 9(c).

11. Suppose all eigenvalues of A have negative real parts and that $x^1(t)$ is a solution of $x' = Ax + F(t)$ which is unbounded as $t \to +\infty$. Let $x^2(t)$ be any other solution. Then $x = x^2 - x^1$ is a solution of the undriven system $x' = Ax$. Since the eigenvalues of A have negative real parts, $\|x(t)\| \to 0$ as $t \to +\infty$. So, $x^2(t) = x^1(t) + x(t)$ is the sum of a vector function which is unbounded and another which decays to 0 as $t \to +\infty$, and $x^2(t)$ is also unbounded as $t \to \infty$.

13. Group project. We know $x' = Ax + F(t)$ where $F(t + T) = F(t)$ has a forced oscillation $x(t)$ of period T if $x(0) = x(T) = e^{TA}\left[x(0) + \int_0^T e^{-sA}F(s)ds\right]$, i.e., if $(I - e^{TA})x(0) = e^{TA}\int_0^T e^{-sA}F(s)ds$. Such an equation is uniquely solvable for $x(0)$ if and only if the matrix $I - e^{TA}$ is nonsingular. So we need to show that $I - e^{TA}$ is nonsingular if $2\pi i/T$ is not an eigenvalue of A. Suppose, to the contrary, that $I - e^{TA}$ is singular. Then for some nonzero vector v, $[I - e^{TA}]v = 0$, i.e., $e^{TA}v = v$. Then $y = e^{tA}v$ is a periodic solution of $y' = Ay$ of period T since the solution returns to its initial state v when $t = T$. But by Theorem 7.6.4 all components of all solutions of $y' = Ay$ are polynomial-exponentials, where the exponential factors have the form $e^{\lambda_j t}$, λ_j an eigenvalue of A. The only way any of these functions could be periodic of period T would be if some $\lambda_j = 2\pi i/T$ (leading to terms such as $\cos(2\pi t/T)$ or $\sin(2\pi t/T)$), but by assumption $2\pi i/T$ is *not* an eigenvalue of A. So the supposition that $I - e^{TA}$ is singular is wrong, and $I - e^{TA}$ is nonsingular. So there is a unique initial state $x(0) = [I - e^{TA}]^{-1}e^{TA}\int_0^T e^{-sA}F(s)ds$ which gives a periodic forced oscillation $x(t)$.

Background Material: Proofs of Root Tests

We have inserted below the proof of the Coefficient Test (Theorem 7.9.6) and the statement of the complete Routh Test (but not its proof) along with two examples. We also provide the proof of the Gerschgorin Disks Theorem (Theorem 7.9.9).

Verification of Theorem 7.9.6 (Coefficient Test).

We will prove the contrapositive, i.e., we will show that *if* all the roots of the polynomial with real coefficients $P(r) = r^n + a_{n-1}r^{n-1} + \cdots + a_0$ have negative real parts, *then* the coefficients of $P(r)$ must all be positive. Suppose the real roots (all negative) are denoted by $-s_1, \ldots, -s_j$ (so $s_1, \ldots, s_j$ are positive) with respective multiplicities $m_1, \ldots, m_j$, and the pairs of complex conjugate roots (all with negative real parts) are $-\alpha_1 \pm i\beta_1$, $\ldots, -\alpha_k \pm i\beta_k$, with respective multiplicities $n_1, \ldots, n_k$. Note that $(r + \alpha_p - i\beta_p)(r + \alpha_p + i\beta_p) = (r^2 + 2\alpha_p r + \alpha_p^2 + \beta_p^2)$. By the Fundamental Theorem of Algebra (Appendix B.4), we see that

$$P(r) = (r + s_1)^{m_1} \cdots (r + s_j)^{m_j}(r^2 + \alpha_1 r + \alpha_1^2 + \beta_1^2)^{n_1} \cdots (r^2 + \alpha_k r + \alpha_k^2 + \beta_k^2)^{n_k} \quad \text{(i)}$$

Since all the s_i's and α_i's are themselves positive numbers, we see that when the right side of (i) is multiplied out, the coefficients of $P(r)$ are all positive. This proves the theorem.

The Routh Test: As the example $r^3 + r^2 + r + 1 = (r + 1)(r^2 + 1) = (r + 1)(r + i)(r - i)$ shows, the positivity of all the coefficients of a polynomial does not necessarily guarantee that all the roots have negative real parts. However, the Routh Test will handle this case, and, in fact, it can be used with any polynomial whose leading coefficient is $+1$.

Routh Test

Suppose that the coefficients of $P(r) = r^n + a_{n-1}r^{n-1} + \cdots + a_0$ are real. Then all roots of $P(r)$ have negative real parts if and only if all entries in the first column of the *Routh Array* are positive:

$$\begin{bmatrix} 1 & a_{n-2} & a_{n-4} & \cdots & a_{n-2p+2} \\ a_{n-1} & a_{n-3} & a_{n-5} & \cdots & a_{n-2p+1} \\ b_{11} & b_{12} & b_{13} & \cdots & 0 \\ b_{21} & b_{22} & b_{23} & \cdots & 0 \\ \vdots & \vdots & \vdots & & \vdots \\ b_{n-1,1} & b_{n-1,2} & b_{n-1,3} & \cdots & 0 \end{bmatrix}$$

where $n = 2p - 2$ or $2p - 1$ [so, $p = (n+2)/2$, or $(n+1)/2$, whichever is an integer, a_0 is the first or second entry of the last column], $a_k = 0$ for $k < 0$, $a_n = 1$, and

$$b_{1j} = \frac{a_{n-1}a_{n-2j} - a_n a_{n-2j-1}}{a_{n-1}}, \quad b_{2j} = \frac{b_{11}a_{n-2j-1} - a_{n-1}b_{1,j+1}}{b_{11}}, \quad j < p$$

$$b_{ij} = \frac{b_{i-1,1}b_{i-2,j+1} - b_{i-2,1}b_{i-1,j+1}}{b_{i-1,1}}, \quad i > 2, \ j < p$$

For example,

$$b_{11} = \frac{a_{n-1}a_{n-2} - 1 \cdot a_{n-3}}{a_{n-1}}, \qquad b_{12} = \frac{a_{n-1}a_{n-4} - 1 \cdot a_{n-5}}{a_{n-1}}$$

$$b_{21} = \frac{b_{11}a_{n-3} - a_{n-1}b_{12}}{b_{11}}, \qquad b_{22} = \frac{b_{11}a_{n-5} - a_{n-1}b_{13}}{b_{11}}$$

$$b_{31} = \frac{b_{21}b_{12} - b_{11}b_{22}}{b_{21}}, \qquad b_{32} = \frac{b_{21}b_{13} - b_{11}b_{23}}{b_{21}}$$

The algorithm for constructing the Routh Array looks formidable, but for any particular polynomial it is easy to use. The rows are constructed from the top down and from left to right. If a zero or a negative number ever appears in the left-hand column of the Routh Array during its construction, there is no point in proceeding further since the criterion has already been violated, and the polynomial must have at least one root with nonnegative real part. The Routh Test need only be used for a polynomial with positive coefficients since the Coefficient Test handles the other cases.

The Routh Array for $r^3 + r^2 + r + 1$ is

$$\begin{bmatrix} 1 & 1 \\ 1 & 1 \\ 0 & - \\ - & - \end{bmatrix}$$

where we haven't bothered to calculate all the entries because the 0 in the first column tells us already that the polynomial has at least one root with non-negative real part. Here's a second example.

The Routh Array for the polynomial, $r^4 + 2r^3 + 3r^2 + 4r + 1/2$ is

$$\begin{bmatrix} 1 & 3 & 1/2 \\ 2 & 4 & 0 \\ 1 & 1/2 & 0 \\ 3 & 0 & 0 \\ 1/2 & 0 & 0 \end{bmatrix}$$

All roots of the given polynomial have negative real parts since all entries in the first column of the Routh Array are positive.

The Array for the polynomial $r^5 + 2r^4 + 3r^3 + 4r^2 + r/2 + 1/4$ is

$$\begin{bmatrix} 1 & 3 & 1/2 \\ 2 & 4 & 1/4 \\ 1 & 3/8 & 0 \\ 15/4 & 1/4 & 0 \\ 37/120 & 0 & 0 \\ 1/4 & 0 & 0 \end{bmatrix}$$

The roots of the polynomial all have negative real parts because the entries in the first column are positive.

Verification of Gerschgorin Theorem 7.9.9:

Suppose that λ is an eigenvalue of $A = [a_{ij}]$, $v = [v_1 \cdots v_n]^T$ a corresponding eigenvector. Since $\lambda v = Av$, we have for each i, $i = 1, \ldots, n$, that

$$\lambda v_i = \sum_{j=1}^{n} a_{ij} v_j, \quad \text{and so,} \quad (\lambda - a_{ii}) v_i = \sum_{j=1, j \neq i}^{n} a_{ij} v_j \tag{ii}$$

Suppose that v_k is the component of v of maximal magnitude; i.e.,

$$|v_k| = \max_{1 \leq r \leq n} |v_r| \neq 0$$

Set $i = k$ in (ii), divide $(\lambda - a_{kk})v_k = \sum_{j=1, j \neq k}^{n} a_{kj} v_j$ by v_k, and take magnitudes:

$$|\lambda - a_{kk}| = \left| \sum_{\substack{j=1 \\ j \neq k}}^{n} a_{kj} \frac{v_j}{v_k} \right| \leq \sum_{\substack{j=1 \\ j \neq k}}^{n} |a_{kj}| \left| \frac{v_j}{v_k} \right| \leq \sum_{\substack{j=1 \\ j \neq k}}^{n} |a_{kj}| = r_k$$

where we have used the Triangle Inequality and the fact that $|v_j/v_k| \leq 1$. So, λ lies in the k-th Gerschgorin row disk. Because A and A^T have the same eigenvalues, λ also lies in at least one Gerschgorin column disk. The collection of all eigenvalues lies in the union of all the row disks and also in the union of all the column disks. Note that this does *not* imply that each disk has one or more eigenvalues in it. The proof of the final conclusion about disjoint unions of disks is omitted.

7.10 Lead Flow, Noise Filter: Steady States

Comments

In this section we return to the lead-level model of Section 7.1. We also introduce a two-loop electrical circuit that is an effective noise filter in a communications channel It is also a low-pass filter, because what it actually does is filter out high-freqencies, and most noise is high-frequency. Now that we have a theory of linear systems we can use it to describe theoretically (as well as by computer simulations) just how the solutions of these systems behave as parameters in the systems are changed. In particular we can use the Routh Test (Theorem 7.9.7) and the Gerschgorin Disks (Theorem 7.9.9) to conclude that the eigenvalues of the system matrix of, respectively, the circuit system and the lead-level system all have negative real parts. This puts these models squarely in the domain of systems with steady-states. And we exploit that by showing how the steady-state periodic output voltage (in the circuit system) changes in amplitude, depending on the frequency of the input voltage. We show how the lead levels in the lead system change with heavier doses of an anti-lead medication.

1. According to text equation (11), the input voltage V_1 of amplitude $|a_0|$ is changed to an output voltage V_2 of amplitude $|a_0[p(i\omega)LRC_1C_2]^{-1}|$. Since

$$-p(i\omega) = (i\omega)^3 + \frac{(i\omega)^2}{RC_1} + \frac{1}{L}\left(\frac{1}{C_1} + \frac{1}{C_2}\right)(i\omega) + \frac{1}{LRC_1C_2}$$

we see that for ω close to 0, $|p(i\omega)| \approx 1/(LRC_1C_2)$, that is, $|LRC_1C_2p(i\omega)| \approx 1$ when ω is close to zero. The low-frequency inputs pass through the circuit with little attenuation. On the other hand if $|\omega|$ is large, then the dominant term in $|p(i\omega)LRC_1C_2|^{-1}$ is $|\omega^3 LRC_1C_2|^{-1}$. For ω large enough this implies that the amplitude of the response is only a tiny fraction of the amplitude of the input voltage, the circuit effectively acting to filter out high frequencies.

3. We see from text equation (12) that the magnitude of $|V_2/V_1|$ is $= [(1 - LC_2\omega^2)^2 + R^2\omega^2(C_1 + C_2 - \omega^2 LC_1C_2)^2]^{-1/2}$, a quantity that we shall denote by $G(\omega)$.

   (a). If $L = 1$, $R = 1$, $C_1 = C_2 = 1$, then

   $$G(\omega) = [1 + 2\omega^2 - 3\omega^4 + \omega^6]^{-1/2}$$

   So, $G(1) = 1$, $G(2) = 1/5$, $G(50) \approx 0$. The frequency 1, 2, and 50 terms are, respectively, passed with no change in amplitude, partially filtered with the amplitude reduced by 80%, and (nearly) stopped.

   (c). If $L = 1$, $R = 100$, $C_1 = C_2 = 0.01$, then

   $$G(\omega) = [1 + \omega^2/50 - 3\omega^4/10^4 + \omega^6/10^6]^{-1/2}$$

   So, $G(1) \approx 0.99$, $G(2) \approx 0.96$, $G(50) \approx 0$, and the frequency 1, 2, and 50 terms are, respectively, passed with little change in amplitude, passed with a slight amplitude reduction, and (nearly) stopped.

5. The system is text system (1).

   (a). If the intake rate is lowered from I_1 to αI_1, where α is a constant between 0 and 1, then the constant steady-state (Theorem 7.9.1) is

   $$\begin{bmatrix} y_1 \\ y_2 \\ y_3 \end{bmatrix} = -A^{-1}\begin{bmatrix} \alpha I_1 \\ 0 \\ 0 \end{bmatrix} = -\alpha A^{-1}\begin{bmatrix} I_1 \\ 0 \\ 0 \end{bmatrix}$$

   The equilibrium levels in the compartments are lowered from the levels of $-A^{-1}[I_1 \ 0 \ 0]^T$ by the same factor α.

   (c). We want to find the smallest value of a, $0 < a < 0.1$ so that if the intake rate I, of lead into the blood is given by $49.3\exp[a(400 - t)]$ for $t > 400$, then the lead level in the bones never exceeds 3000 micrograms. See Fig. 5(c). The solid line corresponds to $a = 0$, while the dashed line corresponds to $a = 0.0102$, which seems to be large enough to keep the equilibrium level less than 3000 milligrams. The value of a is determined experimentally by using a numerical solver.

Problem 5(c).

Background Material: Compartment Models

The lead system is an example of a compartment model in which every compartment is "open to the environment." That is, the environment outside the three compartments of blood, tissues, and bones can be reached by a directed sequence of arrows from each compartment. Let's give a definition of a general compartmental matrix, and then specialize that to an open compartmental matrix, a particular example of which is the lead model coefficient matrix.

The square matrix $A = [a_{ij}]$ of real numbers is a *compartmental matrix* if:

- Every diagonal entry in A is nonpositive.

- Every nondiagonal entry in A is nonnegative.

- The sum of the i-th column entries, $\sum_{j=1}^{n} a_{ji}$, is nonpositive.

Here is why our experience with compartment models in Sections 1.8, 7.1, and 7.10 suggests that these three conditions are reasonable. Thinking of the matrix we would construct for an assemblage of boxes and arrows, we see that the first condition corresponds to the fact that the total rate out of each compartment and into other compartments and the environment must be nonpositive just because it is an *exit* rate. The second condition says that a_{ji}, $j \neq i$, which measures the entrance rate into compartment j from compartment i can't be negative (after all, it is an *entrance* rate). The third condition is less evident. But, recall from the lead model in Section 7.1 that $a_{ii}x_{ii}$ is the negative of the *exit* rate from compartment i, i.e., measures the rate of everything leaving compartment i. So, $a_{ii} = -\sum_{j=1, j \neq i}^{n} a_{ji} - a_{0i}$, where a_{0i} is the exit rate constant from compartment i to the environment. Since a_{0i}, a_{ji} $(j \neq i)$ are all nonnegative, a_{ii} is nonpositive.

We have the following theorem about the eigenvalues of any compartmental matrix. As might be expected from the use of Gerschgorin disks to prove that all eigenvalues of the lead model system matrix (first part of Section 7.10), we validate this theorem by using Gerschgorin disks.

THEOREM

> Compartmental Matrix Theorem. All eigenvalues of a compartmental matrix have nonpositive real parts, and none is pure imaginary.

Here is how we can justify this very important result. As noted above, every diagonal entry a_{ii} is nonpositive and $a_{ii} = -\sum_{j=1, \neq i}^{n} a_{ji} - a_{0i}$. But the a_{ji} terms, $j \neq i$, are nonnegative, and so $\sum_{j=1, \neq i}^{n} a_{ji}$ is just the radius of the i-th Gerschgorin column disk whose center is at $a_{ii} \leq 0$ on the real line in the complex plane. This means that the i-th column disk lies entirely inside the left half of the complex plane, except possibly for the origin, which lies on the edge of the disk if $a_{0i} = 0$. This proves the theorem.

It is also known that if A is a compartmental matrix and $x_i(0) \geq 0$, then $x_i(t) \geq 0$ for all $t \geq 0$, $i = 1, \ldots, n$, if $x(t) = [x_1(t) \ldots x_n(t)]^T$ is a solution of $x' = Ax$.

Now let's define a compartmental system (and matrix) where every compartment is "open to the environment." If for every i there is a directed chain of arrows with positive exit rate constants which leads from compartment i to the external environment, then the system is said to be *open to the environment*. The corresponding compartmental matrix A is *open to the environment* (often called an *open compartmental matrix*) if there is a chain of compartments $j_1, \ldots, j_r$ for which $a_{j_1 i}, a_{j_2 j_1}, \ldots$, and $a_{0 j_r}$ are positive. It isn't hard to verify that the lead matrix is open. We have the following theorem (proof omitted) for any open compartmental matrix:

THEOREM

> Washout Theorem.
>
> If A is an open compartmental matrix, then all its eigenvalues have negative real parts, so all solutions of $x' = Ax$ tend to 0 as $t \to +\infty$.

We would expect the Washout Theorem to be true because the substance being tracked through the system leads out into the environment from each compartment either directly, or indirectly through a chain of other compartments.

All of this can be extended to compartmental systems with "traps," which are compartments or groups of compartments which "absorb," but never disgorge. In a way, the outside environment is a trap, and we could add it to the set of boxes of the model as an additional compartment (the "zero-th" compartment). The enlarged model is no longer open; in fact, it is said to be a *closed compartmental model*. The corresponding matrix A is singular (i.e., 0 is an eigenvalue), and it can be shown that each solution $x(t)$ of $x' = Ax$ tends to some equilibrium point b, where $b_i \geq 0$ if $x_i(0) \geq 0$, $i = 1, \ldots, n$.

Well, that's all we have to say here about compartment models, but here are some sources for further study:

1. *Linear Models in Biology*, M.R. Cullen [Ellis Horwood Limited, Chichester, 1985].

2. *Compartmental Modeling and Tracer Kinetics*, David H. Anderson [v. 50 of *Lecture Notes in Biomathematics*, Springer-Verlag, Berlin].

3. *Compartmental Analysis in Biology and Medicine*, John A. Jacquez [Univ. of Michigan Press, Ann Arbor, 1985 (2nd ed.)].

7.11 General Linear Systems

Comments

We avoided general linear differential systems until this final section because almost all the models we consider in this text lead to linear systems with constant coefficients. The reason for this is that there is a lot of theory for linear systems with nonconstant coefficients, but not many practical methods for finding solution formulas. Similarly, we didn't do much in Chapter 3 either with second-order linear ODEs with nonconstant coefficients (except at the end of the chapter). The one really useful solution formula technique for general second-order linear ODEs is the power series approach described in Chapter 11. If you are going to do that chapter we recommend that you first read Section 3.7.

1. The Fundamental Theorem 7.11.1 says that the IVP $x' = A(t)x$, $x(t_0) = e^j$, has a unique solution $x^j(t)$, for each $j = 1, 2, \ldots, n$. Note that the matrix $[x^1(t) \ \ldots \ x^n(t)]$ evaluated at $t = t_0$ is just the identity matrix I_n. The Wronskian

$$W(t) = \det[x^1(t) \ \ldots \ x^n(t)]$$

has the value at $t = t_0$, $W(t_0) = 1 \neq 0$ and so $\{x^1(t), \ldots, x^n(t)\}$ is a basic solution set.

3. The system is $x' = \begin{bmatrix} t^{-1} & 0 & -1 \\ 0 & 2t^{-1} & -t^{-1} \\ 0 & 0 & t^{-1} \end{bmatrix} x, \ t > 0$.

(a). Direct substitution shows that the columns of $M(t)$ are solutions of the given system. Since $\det M(t)$ is the Wronskian of these three solutions and since $\det M(1) = 2$, it follows that $M(t)$ is a basic solution matrix.

5. **(a).** Let $x(t)$ solve the system $x' = Ax$ and $z(t)$ solve the system $(z^T)' = -z^T A$. Differentiating the scalar product $z^T x$ we obtain

$$(z^T x)' = (z^T)'x + z^T x' = -z^T Ax + z^T Ax = 0$$

for all t. Therefore $z^T x = $ constant, for all t.

7. **(a).** That $X(t) = \begin{bmatrix} t^2 & t \\ 2t & 1 \end{bmatrix}$ is a solution matrix of $x' = y$, $y' = -2x/t^2 + 2y/t$, $t > 0$, follows from a straightforward calculation, i.e., show that $x = t^2$, $y = 2t$ is a solution, and then that $x = t$, $y = 1$ is also a solution.

(c). See Graph 1 for the orbits and Graph 2 for the component curves, given the initial data $x(1) = 1$, $y(1) = -5, -4, \ldots, 4, 5$.

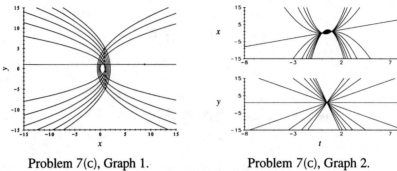

Problem 7(c), Graph 1.   Problem 7(c), Graph 2.

9. Let $x(t) = \Phi(t, t_0)c(t)$ be a solution of $x' = A(t)x + F$, where the "varied parameter" vector $c(t)$ is to be determined. We have $x' = \Phi'c + \Phi c' = A\Phi c + \Phi c' = A\Phi c + F$. So $\Phi c' = F$ and $c'(t) = \Phi^{-1}(t, t_0)F(t)$. We have $c(t) - c(t_0) = \int_{t_0}^{t} \Phi(t_0, s)F(s)\, ds$, where we have used the property that $\Phi^{-1}(t, t_0) = \Phi(t_0, t)$. Setting $c(t_0) = 0$ [which amounts to solving the IVP where $x(t_0) = 0$], we have that $c(t) = \int_{t_0}^{t} \Phi(t_0, s)F(s)\, ds$, and we have constructed a soution of the desired form $x = \Phi(t, t_0)c(t)$.

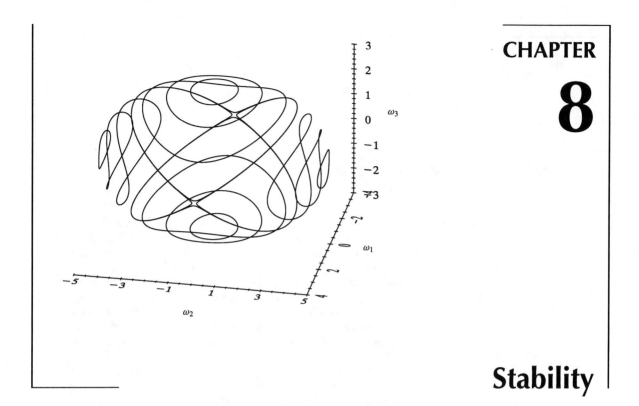

CHAPTER

8

Stability

At the end of the 19th century A. M. Lyapunov introduced a series of tests that could be used to determine the stability properties of a system of ODEs without using solution formulas. These tests involve the use of a scalar function of the state variables and its derivative following the motion. These Lyapunov functions are often based on the total energy of the system, if that can be defined, or else a quadratic form in the state variables. In Section 8.1 we define stability (both neutral and asymptotic) of a system of autonomous ODEs at an equilibrium point and illustrate the definition with many examples. We also give the criterion for stability for an autonomous linear system $x' = Ax$. In Section 8.2 we give the result that an autonomous linear system whose system matrix has only eigenvalues with negative real parts is not only asymptotically stable at the origin, but cannot be destabilized by any higher order perturbation. On the other hand, if at least one of the eigenvalues has positive real part, then the unstable linear system cannot be stabilized by any higher order perturbation. Lyapunov's result (Theorem 8.2.1) is actually phrased as follows: If the Jacobian matrix of $f(x)$ at an equilibrium point p has all eigenvalues with negative real parts, then the system $x' = f(x)$ is asymptotically stable at p. If there is at least one eigenvalue with positive real part, then $x' = f(x)$ is unstable at p. These results form the basis of the long-standing technique of linearizing a nonlinear system at an equilibrium point in order to understand the long-term behavior of the solutions of the nonlinear system near the point. In Section 8.3 we generalize the notion of exactness and the existence of an integral given in Section 1.6 and its problem set to the existence of an integral for certain systems of ODEs, which are said to be conservative. Typically, physical systems for which no energy is lost or gained have model systems of ODEs which are conservative, and this analogy is explored. The chapter cover figure displays an integral surface and several orbits of the conservative system that models the angular velocities of a whirling tennis racket. In Section 8.4 we present Lyapunov's tests and the so-called Lyapunov Method. In the process we characterize the definiteness properties of quadratic forms.

Since the material of this chapter is likely to be new and rather strange for many readers, we give lots of examples.

8.1 Stability and Linear Systems

Comments

In this section we lay out the definitions of stability and its refinements, neutral stability and asymptotic stability, for an autonomous system at an equilibrium point. Since the rest of the chapter is devoted to finding tests for stability and instability, we want to be sure the definitions are understood—and so an entire section mostly on definitions is included. If the system is linear, then Theorem 8.1.1 allows us to test the stability properties simply by inspecting the eigenvalues of the system matrix and, if an eigenvalue λ with zero real part is multiple, determining the dimension of its eigenspace. Asymptotic stability of a system at an equilibrium point requires that the system be an attractor (i.e., all orbits passing near the equilibrium point must approach it as $t \to +\infty$) *and also be stable*. At first glance, one might think that an attracting equilibrium point is automatically stable—but that need not be so (see Problem 6).

"Stability" is an everyday word that has many meanings. In science and engineering it is often used to describe something other than what we define here. For example, in engineering a system is sometimes said to be stable if it responds to bounded inputs with bounded outputs (the "BIBO" of Sections 2.3 and 7.9), but that is quite different from our definition.

1. Since only the orbital portraits are given (not the system itself), we use visual clues to make inferences about the unknown system.

    (a). The system is neutrally stable at the origin. Orbits starting near the origin remain near. The orbits look very much like those for the "center" of the linear system of Example 7.7.5.

    (c). The system is locally asymptotically stable at the equilibrium point $(-1, 0)$ because nearby orbits stay nearby as t increases and, in fact, are attracted to the point. For the same reason the system is also locally asymptotically stable at the equilibrium point $(1, 0)$. But the system is unstable at the equilibrium point at the origin because some nearby orbits leave the neighborhood as t increases and never come back.

3. If the system is linear and all the eigenvalues of the matrix of coefficients have negative real parts, then the system is asymptotically stable at the origin by Theorem 8.1.1. The systems of parts **(a)**, **(e)** are linear, so we need only find the eigenvalues λ of the system matrix. The eigenvalues are calculated by the techniques of Sections 7.4 and 7.9 and are listed below (all have negative real parts). Systems **(c)**, **(g)**, **(i)** are nonlinear and are solved explicitly in order to determine their stability properties.

    (a). The ODE $x' = -x$ is linear; the system "matrix" $[-1]$ has eigenvalue $\lambda = -1$; we have global asymptotic stability at $x = 0$. See Fig. 3(a) for some time-state curves, and the margin for the state line.

(c). Separate variables in the ODE $x' = -\sin x$ and integrate to obtain $\ln|\tan(x/2)| = -t + C_1$. So, $x = 2\arctan(Ce^{-t})$, where C is an arbitrary constant. The values of the arctangent are limited to the range $(-\pi/2, \pi/2)$ since we are only interested in the ODE near $x = 0$; $|x|$ is a decreasing function of t and tends to 0 as t increases to ∞. The ODE is asymptotically stable at $x = 0$ since 0 is an attractor and we can set $\delta = \varepsilon$ in the definition of stability. We have only local asymptotic stability since, for example, equilibrium points such as $x = \pm 2\pi$ are *not* attracted to $x = 0$. See Fig. 3(c) for some time-state curves, and the margin for the state line.

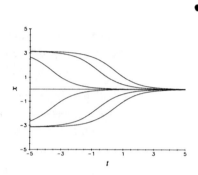

Problem 3(a). Problem 3(c).

(e). The system $x' = x - 3y$, $y' = 4x - 6y$ is linear; the system matrix has eigenvalues $\lambda_1 = -6$, $\lambda_2 = -1$. Both eigenvalues are negative, so the system is globally asymptotically stable at the origin. See Fig. 3(e) for some orbits.

(g). Solving $y' = -5y$, we have $y = c_2 e^{-5t}$, while the nonlinear ODE $x' = -2x^3$ can be solved by separating the variables to obtain $x = c_1(1 + 4c_1^2 t)^{-1/2}$, where c_1 and c_2 are arbitrary constants and $-\infty < t < \infty$ if $c_1 = 0$, and $t > -1/(4c_1^2)$ if $c_1 \neq 0$. The origin $(0, 0)$ is an attractor since $(x(t), y(t)) \to (0, 0)$ as $t \to +\infty$. Moreover, since $\|(x(t), y(t))\|^2 = c_2^2 e^{-10t} + c_1^2(1 + 4c_1^2 t)^{-1}$, we see that $\|(x(t), y(t))\|$ decreases steadily to 0 as t increases to $+\infty$. We may set $\delta = \varepsilon$ in the definition of stability, and the system is globally asymptotically stable at the origin. See Fig. 3(g) for some orbits.

(i). Solving $x' = -x^3$ and $y' = -y^3$ by separating variables, we have that $x = c_1(1 + 2c_1^2 t)^{-1/2}$ and $y = c_2(1 + 2c_2^2 t)^{-1/2}$, where if $c_1, c_2 \neq 0$, then t must be greater than $-1/(2c_1^2)$ and $-1/(2c_2^2)$. Much as in the solution to Problem 1(g), we see that $(0, 0)$ is an attractor since $x(t) \to 0$ and $y(t) \to 0$ as $t \to +\infty$, while $(x(t), y(t))$ decreases steadily to 0 as $t \to +\infty$ [we may set $\delta = \varepsilon$ in the definition of stability]. The system is globally asymptotically stable at the origin. See Fig. 3(i) for some orbits.

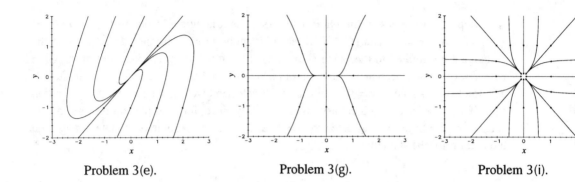

Problem 3(e).   Problem 3(g).   Problem 3(i).

5. System 5(e) is two-dimensional and linear. It is unstable because it has an eigenvalue with positive real part [see Case 3 of Theorem 8.1.1]. Systems 5(a), 5(c), 5(g) are nonlinear. They must be handled by different techniques.

x

$\bullet\ 0$

(a). Separating variables and solving $x' = x^2$, we have $x = x_0/(1 - tx_0)$ if $x(0) = x_0 > 0$. Since $x(t) \to +\infty$ as t increases from 0 to x_0^{-1}, the ODE is unstable at 0. See Fig. 5(a) for some time-state curves, and the margin for the state line.

x

$\bullet\ 0$

(c). Suppose $x > 0$. The solutions of $x' = |x|$ are given by $x = Ce^t$, $C > 0$. Since $x(t) \to +\infty$ as $t \to +\infty$, regardless of the value of C, the ODE is unstable at 0. See Fig. 5(c) for some time-state curves, and the margin for the state line.

(e). The system $x' = 3x - 2y$, $y' = 2x - 2y$ is linear; the system matrix has eigenvalues $\lambda_1 = -1$, $\lambda_2 = 2$. One eigenvalue is positive, so the system is unstable. See Fig. 5(e) for orbits near the unstable origin [Case 3 of Theorem 8.1.1].

(g). The system is $x' = x^3$, $y' = -3y$. We have $y = c_2 e^{-3t}$. Separating variables in $x' = x^3$, and solving for x in terms of t, we have that $x = c_1(1 - 2c_1^2 t)^{-1/2}$, $t < 1/(2c_1^2)$ if $c_1 \neq 0$, and $-\infty < t < \infty$ if $c_1 = 0$. Even though $y(t) \to 0$ as $t \to +\infty$, since $|x(t)| \to \infty$ as t increases from 0 to $1/(2c_1^2)$ (if $c_1 \neq 0$), the system is unstable at the origin. See Fig. 5(g) for some orbits. Note the resemblance to a saddle.

Problem 5(a).

Problem 5(c).

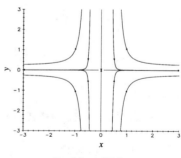

Problem 5(e).   Problem 5(g).

7.   In each part, use Theorem 8.1.1 to determine the stability properties of the system $x' = Ax$,
     where A is the given matrix.

(a). The eigenvalues of the matrix $\begin{bmatrix} -1 & 2 \\ -2 & -1 \end{bmatrix}$ satisfy $(-1-\lambda)(-1-\lambda)+4=0$. Solving, we find $\lambda_1 = \bar{\lambda}_2 = -1 + 2i$. Since each eigenvalue has negative real part, the system
is asymptotically stable.

(c). The eigenvalues of the matrix $\begin{bmatrix} 1 & -1 \\ 2 & -2 \end{bmatrix}$ satisfy $(1-\lambda)(-2-\lambda)+2=0$. Solving,
we find $\lambda_1 = 0$, $\lambda_2 = -1$. Since each eigenvalue is non-positive and λ_1 is zero, the system
is neutrally stable.

(e). The eigenvalues of the matrix $\begin{bmatrix} 5 & 1 \\ 1 & -2 \end{bmatrix}$ satisfy $(5-\lambda)(-2-\lambda)-1=0$. Solving,
we find $\lambda_1 = (3+\sqrt{53})/2$ and $\lambda_2 = (3-\sqrt{53})/2$. Since one of the eigenvalues is positive,
the system is unstable.

(g). By Theorem 7.9.9, the eigenvalues of the matrix $\begin{bmatrix} -10 & 3 & -4 \\ 1 & -12 & 9 \\ -1 & -2 & -5 \end{bmatrix}$ all have negative real parts. To see this, note that each diagonal entry is negative and that on each row
the sum of the magnitudes of the off-diagonal elements has magnitude less than the diagonal entry on that row. So using Gershgorin disks (Section 7.9), the eigenvalues lie entirely
inside the left half of the complex plane, and the system is globally asymptotically stable.

(i). The matrix $\begin{bmatrix} 0 & 1 & 1 \\ 0 & 0 & 1 \\ 0 & 0 & 0 \end{bmatrix}$ has $\lambda = 0$ as a triple eigenvalue. However, the eigenspace
is spanned by the vector $[1 \ 0 \ 0]^T$ and so is of dimension one. So by Theorem 8.1.1, the
system is unstable since the dimension of the eigenspace is less than the multiplicity of the
eigenvalue.

(k). For the matrix $A = \begin{bmatrix} 3 & -2 & 0 & 0 \\ 5 & -3 & 0 & 0 \\ 0 & 0 & 3 & -2 \\ 0 & 0 & 5 & -3 \end{bmatrix}$, the first two equations in the system $x' =$

Ax decouple from the last two. Looking at each pair of equations separately, we find that the eigenvalues $(\pm i)$ have zero real part, so the system here is also neutrally stable.

8.2   Stability of a Nearly Linear System

Comments

Theorem 8.2.1 partly justifies the often heard assertion in science and engineering that the linear terms of a system near an equilibrium point determine the stability properties without having to look at the nonlinearities. It isn't quite that simple, as Problem 4 shows, but the assertion comes close enough to the truth to be widely believed. We outline a part of the proof of Theorem 8.2.1 at the end of Section 8.4 of the Student Resource Manual. We can't do that proof here because it uses some of the ideas of Section 8.4. The Jacobian matrix is introduced here because it often is needed to determine the linearization at an equilibrium point. The table and sketches at the end of the section suggest that the terms node, saddle, and spiral point introduced in Section 7.7 for an equilibrium point of a planar autonomous linear system, still make sense for a nonlinear system. So from now on when (for example) we refer to a saddle point of a system, the system need not be linear, but the Jacobian matrix of the system at the point must have real eigenvalues of opposite sign. In order to make all of this mathematically correct, we introduce the idea of the order of a function at a point and require that the nonlinear terms of a system have order at least 2 and an equilibrium point. **Warning!** Keep your eye out for situations where the Jacobian matrix at an equilibrium point has eigenvalues that are either 0 or pure imaginary. The system may be right on the edge of a stability region and the nonlinear terms can then tip the full system either way—into stability or into instability. Problems 4 and 11 show what can happen.

1. Background for all of the problems here is Theorem 8.2.1.

    (a). The system $x' = -x + y^2$, $y' = -8y + x^2$ is asymptotically stable at the origin since the eigenvalues of the matrix of coefficients of the linear terms at the origin $\begin{bmatrix} -1 & 0 \\ 0 & -8 \end{bmatrix}$ are $-1, -8$ (both are negative).

    (c). The system $x' = 2x - y + x^2 - y^2$, $y' = x - y$ is unstable at the origin since the eigenvalues of the Jacobian matrix $\begin{bmatrix} 2 & -1 \\ 1 & -1 \end{bmatrix}$ at the origin are $(1 \pm \sqrt{5})/2$ (one is positive).

3. As long as no eigenvalue of the Jacobian matrix at an equilibrium point has zero real part, we can tell from the nature of the orbits near an equilibrium point whether the eigenvalues are real and negative, positive, or of opposite signs, or else are complex conjugates with negative or positive real parts.

    (a). See Fig. 3(a) for some orbits of $x' = y$, $y' = -6x - y - 3x^2$. One eigenvalue of the Jacobian matrix at the apparent nonlinear saddle point at the origin must be positive, and the other must be negative. The eigenvalues of the Jacobian matrix at the apparent nonlinear spiral point must be complex conjugates with negative real parts because the orbits spiral into the point as time increases.

(c). See Fig. 3(c) for some orbits of $x' = -x - x^3$, $y' = y + x^2 + y^2$. One eigenvalue of the Jacobian matrix at the apparent nonlinear saddle point must be positive, and the other must be negative. The eigenvalues of the Jacobian matrix at the apparent nonlinear node point must be negative because the orbits approach the point with increasing t.

(e). See Fig. 3(e) for some orbits of $x' = x + xy^2$, $y' = x$. Since the equilibrium points on the y-axis are not isolated, one of the eigenvalues at each equilibrium point must be 0. Since the nonconstant orbits move away from the equilibrium point, the other eigenvalue would seem to be positive. The y-axis through the middle of Fig. 3(e) looks like an earthquake fault line.

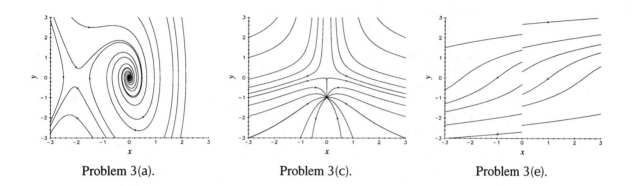

Problem 3(a).   Problem 3(c).   Problem 3(e).

5. The system is $x' = f(x, y) = x - y + x^2 - xy$, $y' = g(x, y) = -y + x^2$.

    (a). From $x - y + x^2 - xy = 0$ and $-y + x^2 = 0$, we have $x = 0$, $y = 0$ or $x = \pm 1$, $y = 1$. This means that the equilibrium points are $(0, 0)$, $(1, 1)$, $(-1, 1)$.

    (c). See figure in the text. The linearized system at the origin is $x' = x - y$, $y' = -y$. The solid arcs are orbits of the nonlinear system, the dashed arcs are orbits of the linear system. Note that the orbits of both systems have common initial points at the upper right and lower left.

7. The system is $x' = y$, $y' = -x/LC - ry/L - g(x, y)/L$. Since $g(x, y)$ is twice continuously differentiable in a neighborhood of $x = 0$, $y = 0$ and is also of order 2 there, $\partial g/\partial x$ and $\partial g/\partial y$ are both 0 at $x = y = 0$. So, g makes no contribution to the Jacobian matrix of the system at the origin. The eigenvalues of the Jacobian matrix at the origin are $\lambda = (-R/L \pm \sqrt{(R/L)^2 - 4/LC})/2$, so have negative real parts. The system is asymptotically stable at the origin, and the values of the nonlinear term g have no effect on the stability property, although g may severely limit the region of asymptotic stability.

9. The nonlinear system is $x' = z + x^2y$, $y' = x - 4y + xz^2$, $z' = \alpha x + 2y - z + x^2$. The linearized version of the system (about the origin) is $x' = z$, $y' = x - 4y$, $z' = \alpha x + 2y - z$. The corresponding characteristic polynomial for the system matrix of this linear system is $\lambda^3 + 5\lambda^2 + (4 - \alpha)\lambda - (2 + 4\alpha) = 0$. The Routh Test (see Section 7.9) for this cubic requires that $4 - \alpha > 0$, $-(2 + 4\alpha) > 0$, $5(4 - \alpha) > -(2 + 4\alpha)$. So we must have $\alpha < 4$,

$\alpha < -1/2$, and $\alpha < 22$; the requirement that α be less than $-1/2$ meets all of the conditions, so the resulting system for these values of α is (locally) asymptotically stable at the origin.

11.   Group project.

8.3 Conservative Systems

Comments

A system of ODEs is conservative if it has a nonconstant integral. The level sets of the integral in state space are unions of orbits, and, so, knowledge of an integral tells us a lot about orbital behavior. What kinds of ODEs have integrals? First-order exact ODEs do (see Problem 10, Section 1.6), and that is why a conservative system may be considered to be a generalization of an exact ODE. A physical system whose total energy when in state $x(t)$ is the same as when in state $x(t_0)$ is conservative; the total energy is the integral. No real physical system has this property, but for many the change in energy as time goes on is negligible and the system can be treated as if it were conservative. Undamped spring systems, undamped pendulums, resistance-free electrical circuits, the Lotka-Volterra predator-prey system, a rotating tennis racket, are all examples of (approximately) conservative systems. We note in this section that conservative systems do *not* have attracting equilibrium points. An integral of a conservative system has some of the properties of the Lyapunov functions to be introduced in Section 8.4. An integral remains constant on each orbit; any orbit touching a level set of an integral remains on the level set, and there need not be any equilibrium points at all. The reason we put this section in this chapter is that (as noted before) the construction of an integral (if there is one) gives information about orbits without having to find solution formulas.

Experiment

Problem 8. Toss a book (not this one) into the air and try to get it to rotate around a body axis. You will soon find that you can get it to rotate about the shortest axis and also the longest axis, but not about the axis of medium length. This leads to a neat analysis of the rate equations for the components of the angular velocity vector, the construction of an integral, and the detection of stable and unstable modes of rotation. The chapter cover picture is connected with this problem.

---

1.   **(a).**   Divide the second equation of the system $x' = 3x$, $y' = -y$ by the first to obtain $dy/dx = -y/3x$. Separating the variables and solving, we have $3\ln|y| = -\ln|x| + C_1$ or $y^3x = C$, where $C = \pm e^{C_1}$. The equation $y^3x = 0$ also defines orbits [i.e., the orbits $x = 0$ and $y = 0$]; we have the integral $K(x, y) = y^3x$, whose level sets define the orbits. The original system is linear with a single equilibrium point at $(0, 0)$. Since the eigenvalues of the system matrix are 3 and -1, the origin is an unstable saddle point. See Fig. 1(a).

(c).   The system is $x' = y^3$, $y' = -x^3$. We obtain an integral by solving $dy/dx = -x^3/y^3$, or $y^3dy + x^3dx = 0$. $K(x, y) = \frac{1}{4}(x^4 + y^4)$ [or any nonzero constant multiple of K] is an integral, and the orbits are the "rounded-off" rectangles defined by $K(x, y) = C$ (neutral stability at the origin). See Fig. 1(c).

Problem 1(a).

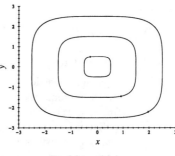
Problem 1(c).

3. The system is $x' = g(y)$, $y' = f(x)$, $z' = 0$. We have $dy/dx = f(x)/g(y)$. Separating variables and integrating, we obtain the integral $K(x, y) = F(x) - G(y)$, where F and G are any antiderivatives of $f(x)$ and $g(y)$, respectively. Note that $z(t) = C$ for all t. The system is conservative in $\mathbb{R}^3$. By Theorem 8.3.2 in this section, the system does not have an attracting equilibrium point in $\mathbb{R}^3$.

5. The system is $x' = Ax$, where $A^T = -A$, that is, A is skew-symmetric.

(a). Let $K(x) = x^T x$. Then, $K' = (x^T)'x + x^T x' = (Ax)^T x + x^T Ax = x^T A^T x + x^T Ax = 0$ since $A^T = -A$. K remains constant along each orbit of $x' = Ax$. Since $K(x)$ is the square of the distance from the point x to the origin, K is not constant on any ball, so $K(x)$ is an integral, and the system is conservative. The level sets $K(x) = C$, C any positive constant, are the "spheres" in $\mathbb{R}^n$ of radius $\sqrt{C}$ centered at the origin: $K(x) = x^T x = x_1^2 + \cdots + x_n^2 = C$. Each orbit lies entirely on one of these spheres, and the origin is not an attractor. The system is neutrally stable at the origin with $\delta = \varepsilon$ in the definition of stability.

7. Group project. See Fig. 7 for some orbits of $x' = y^2 + e^x \cos y + 2 \cos x$, $y' = -e^x \sin y + 2y \sin x$.

Problem 7.

8.4 Lyapunov Functions

Comments

Lyapunov functions provide an alternative to the ε-δ arguments of Section 8.1 to determine the stability properties of a system of ODEs at an equilibrium point. These real-valued functions of the state variables tell us a lot about what is going on in the system, but without having to find solution formulas. Given a system of ODEs with an equilibrium point, it is always possible (but *not* always easy) to find a Lyapunov function whose behavior reveals whether the system is stable, asymptotically stable, or unstable at the equilibrium point. Lyapunov functions are never unique, and this is both an advantage and a difficulty, the former because it allows a lot of flexibility, the latter because there may be no clue about how to construct a Lyapunov function. Although total energy is commonly used as a Lyapunov function for models of physical systems, quadratic forms in the state variables are usually easier to use. In this section we emphasize quadratic forms. There are many other ways to construct Lyapunov functions. Building suitable Lyapunov functions has become a cottage industry for scientists and engineers involved in designing complex systems whose stability may be a matter of life or death, systems ranging from nuclear reactors to the control systems for jets and spacecraft.

1. Since the V-functions are never unique, you may have functions that differ from those given here. However if the functions V and V' given here have certain definiteness properties, your functions must have the *same* properties. This is also the case in the other problems in this set.

   (a). The system is $x' = -4y - x^3$, $y' = 3x - y^3$. If $V = ax^2 + cy^2$, then $V' = 2ax(-4y - x^3) + 2cy(3x - y^3) = (-8a + 6c)xy - 2ax^4 - 2cy^4$. If a and c are chosen as positive constants for which $-8a + 6c = 0$ (e.g., $a = 3$, $c = 4$), then $V = 3x^2 + 4y^2$ is positive definite, and $V' = -2ax^4 - 2cy^4$ is negative definite. The system is asymptotically stable at the origin.

   (c). The system is $x' = -x + 3y$, $y' = -3x - y$. If $V = ax^2 + cy^2$, $V' = 2axx' + 2cyy' = -2ax^2 + 6axy - 6cxy - 2cy^2 = -2x^2 - 2y^2$ if we set $a = c = 1$. Since $V = x^2 + y^2$ is positive definite and $V' = -2x^2 - 2y^2$ is negative definite, the system is asymptotically stable at the origin.

   (e). The system is $x' = (-x + y)(x^2 + y^2)$, $y' = (x + y)(x^2 + y^2)$. With $V = ax^2 + cy^2$, we have $V' = (x^2 + y^2)(-2ax^2 + 2(a - c)xy - 2cy^2) = -2(x^2 + y^2)^2$ if we set $a = c = 1$. Since V is positive definite and V' is negative definite, the system is asymptotically stable at the origin.

   (g). The system is $x' = -x^3 + x^3y - x^5$, $y' = y + y^3 + x^4$. If $V = ax^2 + cy^2$, $V' = 2axx' + 2cyy' = -2ax^4(1 + x^2) + 2cy^2(1 + y^2) + 2(c + a)x^4y$. If $a = -1$ and $c = 1$, then $V = -x^2 + y^2$ is indefinite, while $V' = 2x^4(1 + x^2) + 2y^2(1 + y^2)$ is positive definite; the system is unstable at the origin.

3. **(a).** The system is $x' = -x - x^3$, $y' = -y$. If $V = x^2 + y^2$, then $V' = 2x(-x - x^3) + 2y(-y) = -2x^2 - 2y^2 - 2x^4$. Since V is positive definite and V' is negative definite, the

system is asymptotically stable at the origin.

(c). The system is $x' = \varepsilon x - y + x^2$, $y' = x + \varepsilon y + x^2$, $\varepsilon > 0$. If $V = x^2 + y^2$, then V is positive definite and $V' = 2x(\varepsilon x - y + x^2) + 2y(x + \varepsilon y + x^2) = 2\varepsilon(x^2 + y^2) + 2x^2(x + y)$ which is also positive definite since $\varepsilon > 0$ if $|x|$ and $|y|$ are small enough (say, $|x|$ and $|y|$ less than $\varepsilon/10$); the system is unstable at the origin.

5. The system is $x' = (x - y)^2(-x + y)$, $y' = (x - y)^2(-x - y)$. Every point on the line $y = x$ is an equilibrium point; so the origin cannot be asymptotically stable since it is not an attractor. If $V = x^2 + y^2$, then $V' = 2xx' + 2yy' = -2(x - y)^2(x^2 + y^2) \le 0$; i.e., V' is negative semidefinite, while V is positive definite. The system is neutrally stable at the origin since it is stable, but not asymptotically stable.

7. The system is $x' = (a - by)x$, $y' = (-c + dx)y$, where a, b, c, d are positive constants. The function Q is given by $(y^a e^{-by})(x^c e^{-dx})$. It was shown in Problem 6**(a)** in Section 5.4 that Q attains a strict local maximum at the equilibrium point $P(c/d, a/b)$. So $V = Q(c/d, a/b) - Q(x, y)$ is positive definite in a region containing P. Moreover, $V' = 0$ for all (x, y) in the first quadrant since $Q(x, y)$ is an integral of the equivalent first-order ODE, $(dxy - cy)\, dx + (-ax + bxy)\, dy = 0$. Therefore, the system is stable. Since the orbits are shown in Problem 6, Section 5.4 to be cycles, the system is neutrally stable, and *not* asymptotically stable.

9. The system is $x' = y - xf(x, y)$, $y' = -x - yf(x, y)$. Take $V(x, y) = x^2 + y^2$, which is positive definite. The derivative of V following the motion of the system is $V' = -2(x^2 + y^2)f(x, y)$. The conclusion in each part follows from the fact that $-2(x^2 + y^2)$ is negative definite, so the definiteness properties of V' are opposite to those of f.

(a). Since f is positive semidefinite, V' is negative semidefinite and the system is stable at the origin.

(c). Since f is negative definite, V' is positive definite, and the system is unstable at the origin.

Background Material: Verification of Part 1 of Theorem 8.2.1

We want to show that the system $x' = Ax + P(x)$ is asymptotically stable at the equilibrium point 0 if the eigenvalues of the system matrix A have negative real parts and the "perturbation" $P(x)$ has order at least 2 at $x = 0$. We use the following fact (known as Lyapunov's Lemma) which we don't prove:

> If the eigenvalues of A have negative real parts then the matrix equation
>
> $$A^T B + BA = -I \tag{i}$$
>
> has a matrix solution B such that the scalar function $V(x) = x^T Bx$ is positive definite.

Now we will show that $V(x)$ is a strong Lyapunov function for $x' = Ax + P(x)$ in a neighborhood of $x = 0$. By Lyapunov's Lemma we know that V is positive definite. We will show that V' is negative definite, so by Theorem 8.4.1 (Lyapunov's First Theorem) that means that the system $x' = Ax + P(x)$ is asymptotically stable at the origin. First note that because $P(x)$ has order at least 2 at the origin, then (for some positive constant c) $||P(x)|| \le c||x||^2$ for all x in a neighborhoood of the origin.

Now let's calculate the derivative of $V(x) = x^T Bx$ following the motion of $x' = Ax + P$:

$$V' = (x')^T Bx + x^t Bx' = (x^T A^T + P^T)Bx + x^T B(Ax + P)$$

$$= x^T (A^T B + BA)x + P^T Bx + x^T BP = x^T(-I)x + g(x) \qquad \text{(ii)}$$

$$= -||x||^2 + g(x)$$

where $g(x)$ is the scalar function $P^T Bx + x^T BP$ and where we have used (i). We can use the Triangle Inequality, the Cauchy-Schwarz Inequality, properties of the matrix norms, and the estimate on $||P(x)||$ to estimate $|g(x)|$:

$$|g(x)| = |P^T Bx + x^T BP| \le |P^T Bx| + |x^T BP| = 2|P^T Bx|$$

$$\le 2||P|| \cdot ||Bx|| \le 2c||x||^2 ||Bx|| \qquad \text{(iii)}$$

$$\le 2c||x||^2 ||B|| \cdot ||x|| \le C||x||^3$$

where C is a positive constant. An upper bound for V' is found from (ii) and (iii):

$$V'(x) \le -||x||^2 [1 - C||x||] \qquad \text{(iv)}$$

If x is restricted to the open ball centered at $x = 0$ defined by $|x| < 1/C$, then (iv) implies that V' is negative definite on the open ball. $V(x)$ is a (local) strong Lyapunov function at the origin for the system $x' = Ax + P$, so that system is (locally) asymptotically stable. Note that $V(x)$ also is a (global) strong Lyapunov function for the linear system $x' = Ax$.

Set $u = x - p$ if the equilibrium point is at $x = p$ and $x' = A(x - p) + P(x)$, where $P(p) = 0$ and $P(x)$ has order at least 2 at $x = p$. Then replace the system by $u' = Au + Q(u)$ where $Q(u) = P(u + p)$, and apply the above argument.

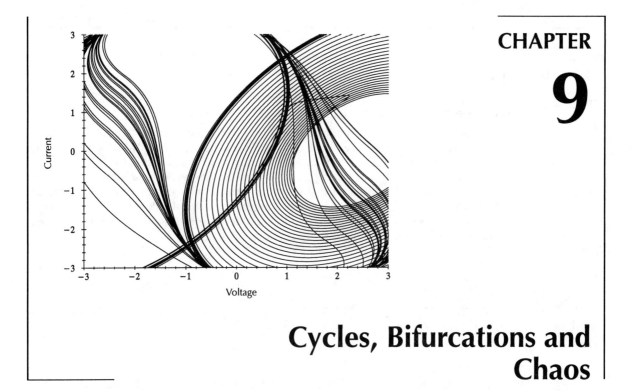

Cycles, Bifurcations and Chaos

Cycles, limit cycles, cycle-graphs, the long-term behavior of a bounded orbit of a planar autonomous system, bifurcations (including the Hopf bifurcation), and the Lorenz system and chaos are the subject of this chapter. We prove a few things, state without proof several basic results, and rely on pictures to help readers understand the ideas. Everything is carried out with just two state variables (usually) and autonomous systems (always) in the first three sections, and this is for two reasons: (1) solution and orbital behavior is easier to see, and (2) Theorem 9.2.1, which gives a precise description of the asymptotic behavior of planar orbits, fails miserably if there are more than two state variables or if the system is nonautonomous. The Lorenz system (Section 9.4) is autonomous, has three state variables, and displays apparently chaotic behavior. Some of the most interesting nonlinear systems appear in the group problems of Sections 9.2–9.4. Any one of these would make an excellent group project. The book and chapter cover figures show the apparently chaotic (but organized) wanderings of voltage and current in the nonlinear scroll circuit (see Section 9.3).

The apparently chaotic dynamics of the scroll circuit of Section 9.3 and the Lorenz system of Section 9.4 raise deep (and as yet unanswered) questions about just what chaos is. This is a wide-open area of intellectual and practical research. We expect that in a few years our approach will seem quaint and a bit old-fashioned, but that often happens when ideas are still in the development stage.

9.1 Cycles

Comments

Cyclical behavior is all around us, and it is not hard to pick up on the idea of a periodic solution as a nonconstant equilibrium state. Limit cycles are a little harder to grasp, which is why we give several examples early on in the section. Since circular cycles are most easily described in polar coordinates, there are a lot of r's and θ's here. The van der Pol system is derived as an extension from a simple linear RLC circuit system to a circuit system with a nonlinear resistor (e.g., a diode). We introduce these circuit models as systems (rather than as second-order ODEs) because that is the most natural way to do the modeling. It is worth emphasizing that the van der Pol cycles are isolated, attracting, and with periods that change with the parameter.

1. The equilibrium points of these systems in r and θ are obtained by finding the points where r' and θ' are zero. If r' is zero when r is zero, then the origin in the xy-plane is an equilibrium point, regardless of the value of θ'. Elsewhere, both r' and θ' must be zero at an xy-equilibrium point. The signs of r' and θ' near a cycle or an equilibrium point will help you decide whether the cycle or point attracts or repels. The figures show arcs of orbits, with arrowheads indicating the direction of time's increase.

(a). The system is $r' = 4r(4-r)(5-r)$, $\theta' = 1$. The equilibrium point is $(0,0)$, which is unstable since r' is positive for $0 < r < 1$. This system has the attracting limit cycle $r = 4$ (r' is positive for $1 < r < 4$ and negative for $4 < r < 5$) and the repelling limit cycle $r = 5$ (r' is positive for $r > 5$). See Fig. 1(a).

(c). The system is $r' = r(1-r^2)(4-r^2)$, $\theta' = 1 - r^2$. The equilibrium points are the origin (unstable since r' is positive for $0 < r < 1$) and all points on the circle $r = 1$. Since $r' > 0$ for $0 < r < 1$ and $r' < 0$ for $1 < r < 2$, the circle $r = 1$ of equilibrium points attracts. In fact, each equilibrium point on the circle is stable. Here's why. Note that $dr/d\theta = r(4-r^2)$. Separating variables and using partial fractions, we have (after a calculation which we omit) $\theta = (1/8) \ln |r^2/(4-r^2)| + C$. So as $r \to 1$, $\theta \to (1/8) \ln(1/3) + C$; that is, by choosing C appropriately, $\theta \to$ any given number θ_0. So the circle $r = 1$ consists of equilibrium points, each of which is approached by two orbits, one with $r \to 1^+$, the other with $r \to 1^-$. So these equilibrium points are neutrally stable. This system has the repelling limit cycle $r = 2$ ($r' > 0$ for $r > 2$ and $r' < 0$ for $1 < r < 2$). See Fig. 1(c) where the circle of neutrally stable equilibrium points is dashed.

(e). The system is $r' = r \cos \pi r$, $\theta' = 1$. The equilibrium point is $(0,0)$ which is unstable since r' is positive near $r = 0$. For $n = 0, 1, 2, \ldots$, this system has the attracting limit cycles $r = 2n + 1/2$ because $r' > 0$ for r near but less than $2n + 1/2$, and $r' < 0$ for r near but larger than $2n + 1/2$; similarly, the limit cycles $r = 2n + 3/2$ are repelling. See Fig. 1(e).

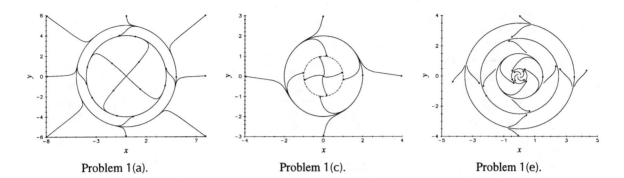

Problem 1(a).   Problem 1(c).   Problem 1(e).

3. The system $r' = r^3 \sin(1/r)$, $\theta' = 1$ has the repelling limit cycles $r = 1/(2n+1)\pi$ $n = 0, 1, \ldots$ and the attracting limit cycles $r = 1/(2n\pi)$, $n = 1, 2, \ldots$. Because $r = 1/(2n+1)\pi \to 0$ and $r = 1/(2n\pi) \to 0$ as $n \to \infty$, the origin is said to be a center-spiral point for this system; the origin is surrounded by a family of cycles separated from one another by annular regions of alternately inward and outward spiraling orbits, because r' changes sign across each cycle.

5. The system $r' = r(1 - r^2)^2(4 - r^2)(9 - r^2)$, $\theta' = 1$ has a repelling limit cycle, $r = 3$, and an attracting limit cycle, $r = 2$; $r = 1$ is a semistable cycle, attracting interior orbits and repelling exterior orbits. This follows because $r' > 0$ if $0 < r < 1$, $1 < r < 2$, or $r > 3$, while $r' < 0$ if $2 < r < 3$. See Fig. 5.

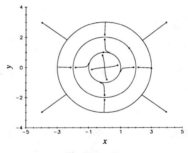

Problem 5.

7. The original system in r, θ, z coordinates is $r' = r(1 - r^2)$, $\theta' = 25$, $z' = \alpha z$. The system in xyz-coordinates is $x' = x - 25y - x(x^2 + y^2)$, $y' = 25x + y - y(x^2 + y^2)$, $z' = \alpha z$. Viewed solely from the xy-plane, the unit circle is an attracting limit cycle. It is an attracting limit cycle in $\mathbb{R}^3$ as well if $\alpha < 0$. See Fig. 7, Graph 1 (where $\alpha = -0.5$) for two nonconstant orbits, one with $x_0^2 + y_0^2 < 1$ and the other with $x_0^2 + y_0^2 > 1$. If $\alpha = 0$, then z is constant and orbits lie in planes $z = z_0$. The unit circle in each plane is a limit cycle that attracts each nonconstant orbit in the plane. However, the cycle $r = 1$, $z = 0$ is *not* an attractor because it does not attract orbits from other planes. See Fig. 7, Graph 2 for orbits in the planes $z = 10$, 40. If $\alpha > 0$ and $x = x(t)$, $y = y(t)$, $z = z(t)$ defines a nonconstant solution, then $x^2(t) + y^2(t) \to 1$ as $t \to +\infty$, while $z(t) \to +\infty$ if $z(0) > 0$. Figure 7, Graph 3 shows the orbit of $x = y = 0$, $z(t) = 0.01e^t$ and two other orbits with $z_0 > 0$, one with $x_0^2 + y_0^2 > 1$ and the other with $x_0^2 + y_0^2 < 1$.

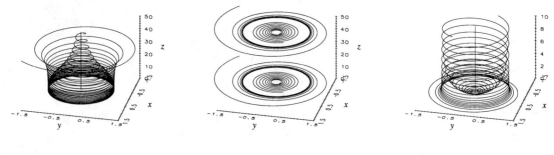

Problem 7, Graph 1.   Problem 7, Graph 2.   Problem 7, Graph 3.

9.   In each figure, orbits in the xy-plane are at the left and x and y component graphs at upper and lower right, respectively.

(a). The system is $x' = y - \mu(x^3 - x)$, $y' = -x$. See Figs. 9(a) for $\mu = 0.1, 5, 10$. The periods appear to be about 7, 10, and 20, respectively. The period appears to increase as μ increases. The x-amplitude seems to remain constant, while the y-amplitude increases as μ increases.

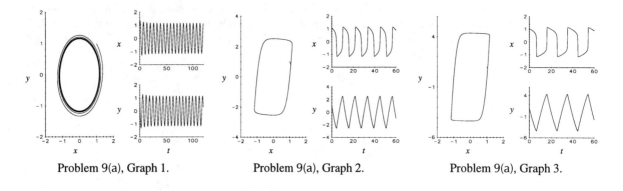

Problem 9(a), Graph 1.   Problem 9(a), Graph 2.   Problem 9(a), Graph 3.

Background Material: Sketch of the van der Pol Cycle Theorem (Theorem 9.1.1)

The form of the rate equation of the van der Pol system

$$x' = y - \mu f(x), \quad y' = -x \tag{i}$$

and the given conditions on μ and f imply that:

- If $x = x(t)$, $y = y(t)$ is a solution of (i), so also is $x = -x(t)$, $y = -y(t)$. Consequently, orbits of (i) in the left half-plane are simply the reflections through the origin of orbits in the right half-plane. For this reason, we only need to carry out our analysis in the right half-plane.

- If there is a cycle, it must enclose the equilibrium point at the origin.

- As noted in Section 5.2, Sign Analysis shows that each orbit Γ of (i) in the right half-plane and above the graph K of $y = \mu f(x)$ moves to the right from initial point A and falls as time increases. Such an orbit crosses K in the vertical downward direction and as τ increases moves to the left and intersects the negative y-axis at a point G. Let D be the point on K above the point $(\beta, 0)$ where $\beta > a$. Sketch a figure in the xy-plane and label the curve K

and the points A, G, and D. The van der Pol graph shown uses $\mu = 1/4$, $f(x) = x^3/3 - x$. Use this graph to locate the points A, B, C, D, E, F, G as well as $x = a = \sqrt{3}$ and $x = \beta$ mentioned in the construction.

- According to the reflection property, Γ is a cycle if and only if the distance from O (the origin) to A equals the distance from O to G.

- If Γ is a cycle when $\beta = \beta_0$ then that cycle is unique and attracting if and only if for $\beta > \beta_0$ the distance $|OA|$ from O to A exceeds $|OG|$, while for $0 < \beta < a$ we have $|OA| < |OG|$. This follows (but not very easily) from the reflection properties of orbits and the fact that distinct orbits cannot intersect.

These properties are used below.

Set $E = x^2/2 + y^2/2$; we see that $dE/dt = x\,dx/dt + y\,dy/dt = x(y - \mu f) + y(-x) = -\mu x f$, Since $dy/dt = -x$, so $-x\,dt = dy$ and we have along Γ

$$dE = \mu f(x)\,dy$$

If we set $\delta E = E_G - E_A$, the difference between the values of E at G and at A, then δE is a well-defined, continuous function of β, the x-coordinate of the point D on Γ. We have that

$$\delta E(\beta) = \frac{1}{2}|OG|^2 - \frac{1}{2}|OA|^2 = \oint_{AG} dE = \oint_{AG} \mu f(x)\,dy$$

The theorem is proven if we show that:

- $\delta E(\beta)$ is positive and bounded for $0 < \beta < a$,

- $\delta E(\beta)$ is strictly decreasing for $\beta \geq a$,

- $\delta E(\beta) \to -\infty$ as $\beta \to +\infty$.

These three properties and the continuity of $\delta E(\beta)$ imply that there is exactly one point β_0 for which $\delta E = 0$. From the formula for $\delta E(\beta)$ and the reflection property for orbits, we see that the orbit Γ corresponding to β_0 must be a cycle since $|OG| = |OA|$. Moreover, the cycle is attracting because the sign of the values of $\delta E(\beta)$ changes from $+$ to $-$ as β increases through β_0. So, we need only prove the three properties of δE listed above.

First, suppose that $0 < \beta < a$. Along the arc of Γ from A to B we have that both $f(x)$ and dy are negative because in this case the arc lies entirely in the vertical strip $0 < x < a$. Since $\mu > 0$ we have that $\delta E(\beta) > 0$. Because $\delta E(\beta)$ is continuous for all $\beta \geq 0$, the Maximum Value Theorem (Theorem B.5.1) implies that $\delta E(\beta) > 0$ has a positive upper bound E_{max} on the interval $0 \leq \beta \leq a$.

Second, suppose that $\beta > a$, and divide the line integral in the formula for $\delta E(\beta)$ into three parts:

$$\delta E(\beta) = \oint_A^G \mu f\,dy = \oint_A^B + \oint_F^G + \oint_B^F = L_1(\beta) + L_2(\beta)$$

where L_1 is the sum of the line integrals from A to B (B is the point on the orbit above $x = a$, and F is the point on the orbit below $x = a$) and from F to G, and $L_2(\beta)$ is the line integral from B to F. We shall show that $L_1(\beta)$ is a positive and decreasing function of $\beta \geq a$ and that $L_2(\beta)$ is negative and decreasing on the same interval. $L_1(\beta)$ is positive because μf and dy are negative (except possibly at A, B, F, and G) on the arcs of Γ involved in calculating the value of $L_1(\beta)$. $L_1(\beta)$ decreases as β increases because

$$\mu f\,dy = \mu f \frac{dy}{dx}dx = -\frac{x\,dx}{y - \mu x f(x)}$$

and, as β increases, the arcs of Γ from A to B and from F to G move outward. This means that as β increases the denominator $y - \mu f(x)$ increases for each fixed x, $0 < x < a$, and the quotient decreases, as must $L_1(\beta)$.

Turning to the function $L_2(\beta)$, $\beta > a$, we see that L_2 is negative because dy is negative on the arc of Γ from B to F while $\mu f(x)$ is positive (except at B and F). Finally, $L_2(\beta)$ is a decreasing function of β, $\beta \geq a$, because dy is negative (as before) and $\mu f(x)$ is positive and increasing for each x, $x \geq a$. So $\delta E(\beta)$ decreases from a positive value E_0 at $\beta = a$ as β increases, $\beta \geq a$.

$$x' = y - 0.25(x^3/3 - x), \qquad y' = -x$$

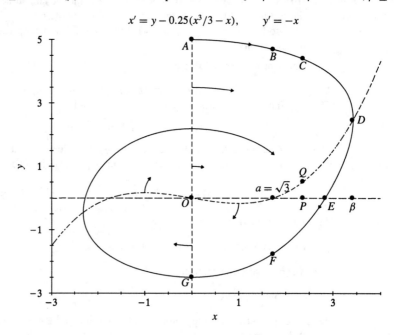

Graphs of $y = \mu f(x)$ and of van der Pol orbits.

Third, to show that $\delta E(\beta) \leq E_0$, we only need show that $L_2(\beta) \rightarrow -\infty$ as $\beta \rightarrow +\infty$. Fix the point P on the x-axis as shown in the figure. We have that

$$L_2(\beta) = \oint_B^F < \oint_C^E \mu f(x)\, dy \leq -|PQ| \cdot |PC|$$

because $\mu F\, dy$ is everywhere negative along the arc from B to F (except at B and F) and the magnitude of a line integral is at least as large as the product of the minimal magnitude of the integrand μf ($|PQ|$ in this case) along the arc from C to E and the arc length (which is at least $|PC|$ here). But $|PC| \rightarrow \infty$ as $\beta \rightarrow +\infty$. So, the above inequality for $L_2(\beta)$ implies that $L_2(\beta) \rightarrow -\infty$, and we see that the van der Pol system must have a cycle.

9.2 Long-Term Behavior

Comments

The nature of the long-term behavior of a bounded solution of an autonomous planar system is the core of this section, but that only makes sense after the idea of a rather mysterious object called a cycle-graph has been explored. This is why we have several examples of cycle-graphs in the text and problem set. The reason for the emphasis on limit sets is that for applications in engineering and science one is often more interested in the long-term behavior of solutions than in the short-term behavior, and the limit sets show the asymptotic nature of orbits. The main result is Theorem 9.2.1 which says that a bounded orbit must tend to an equilibrium point, or to a cycle, or to a cycle-graph, and there are no other possibilities. In other words, chaotic motion *cannot* happen in this setting of a planar autonomous system. This is in marked contrast to what we will see in Section 9.4, and to some extent in Section 9.3. Bendixson's Negative Criterion (Theorem 9.2.4) is a little gem; it is useful and has an elegant proof that uses Green's theorem (see the end of this section). In doing Problem 14 people have created marvelous ODE drawings ranging from the classical kitty shown in the margin to something that might have been done by Picasso in one of his more avant-garde phases.

1. **(a).** The system is $x' = y$, $y' = -x$. Divide y' by x' to obtain $dy/dx = -x/y$. So, $x^2 + y^2 = C^2$ defines a circular orbit (i.e., a cycle) about the origin (which is then neutrally stable) for each value of C. Furthermore, the general solution of the original system is $x = C\cos(t + \phi)$ and $y = -C\sin(t + \phi)$, where C and ϕ are arbitrary constants. Consequently, the solutions through the points $(0,0)$ and $(1,1)$ are, respectively, $x = 0$, $y = 0$, and $x = \cos t$, $y = -\sin t$. The negative and the positive limit sets of the orbit Γ through $x = 0$, $y = 0$ are the origin, which is the only equilibrium point; so $\alpha(\Gamma) = \omega(\Gamma) = \Gamma = (0, 0)$. The negative and the positive limit sets of the second orbit are the second orbit itself, which is a cycle; so $\alpha(\Gamma) = \omega(\Gamma) = \Gamma = $ unit circle. There are no cycle-graphs.

    (c). The van der Pol system is $x' = y - (x^3/3 - x)$, $y' = -x$. The negative and the positive limit sets of the orbit Γ through $(0.1, 0)$ are, respectively, the origin and the cycle shown in Fig. 1(c); so $\alpha(\Gamma)$ is the origin and $\omega(\Gamma)$ is the cycle. For the orbit though $(3, 0)$, the negative and positive limit sets are, respectively, the empty set (because the orbit becomes unbounded as t decreases) and the van der Pol cycle shown. The origin is the only equilibrium point, and the cycle shown in Fig. 1(c) is the only cycle. There are no cycle-graphs.

    (e). The (polar) system is $r' = r(1 - r^2)(4 - r^2)$, $\theta' = 5$. The negative and the positive limit sets of the orbit through $r = 1/2$, $\theta = 0$ are, respectively, the origin and the cycle $r = 1$, because r' is positive for $0 < r < 1$ and $r' = 0$, $\theta' \neq 0$ on $r = 1$. The negative and the positive limit sets of the orbit through $r = 3/2$, $\theta = 0$ are, respectively, the cycles $r = 2$ and $r = 1$, because $r' < 0$ for $1 < r < 2$. The negative and the positive limit sets of the orbit through $r = 5/2$, $\theta = 0$ are, respectively, the cycle $r = 2$ and the empty set (because $r(t) \to +\infty$ as t increases) since $r' > 0$ for $r > 2$. These results follow from determining the sign of r' in the regions $0 < r < 1$, $1 < r < 2$, $r > 2$. There are no cycle-graphs.

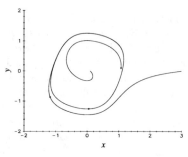

Problem 1(c).

3.   Each system has the form $x' = f(x, y)$, $y' = g(x, y)$. We let the symbol Δ denote the quantity $\partial f/\partial x + \partial g/\partial y$ given in Theorem 9.2.4.

    (a). The system is $x' = 2x - y + 36x^3 - 15y^2$, $y' = x + 2y + x^2 y + y^5$. Here Δ is $\partial(2x - y + 36x^3 - 15y^2)/\partial x + \partial(x + 2y + x^2 y + y^5)/\partial y = 4 + 109x^2 + 5y^4 > 0$. According to Bendixson's Negative Criterion, there is no cycle or cycle-graph anywhere in the xy-plane.

    (c). The system is $x' = x - xy^2 + y \sin y$, $y' = 3y - x^2 y + e^x \sin x$, and Δ is $\partial(x - xy^2 + y \sin y)/\partial x + \partial(3y - x^2 y + e^x \sin x)/\partial y = 4 - (x^2 + y^2) > 0$ if $x^2 + y^2 < 4$. According to Bendixson's Negative Criterion, there is no cycle or cycle-graph anywhere in the disk $x^2 + y^2 < 4$.

5.   **(a).** The system is $x' = e^x + y^2$, $y' = xy$. Since x' is always positive, no orbit can return to its starting point as t increases, and there are no cycles.

    (c). The (polar) system is $r' = r \sin r^2$, $\theta' = 1$. Note that when $r = \sqrt{n\pi}$, $n = 0, 1, \ldots$, then $r' = 0$. Also, $r' > 0$ for $2k\pi < r^2 < (2k+1)\pi$ and $r' < 0$ for $(2k+1)\pi < r < (2k+2)\pi$, $k = 0, 1, \ldots$. So $r^2 = 2k\pi$ defines repelling limit cycles, and $r^2 = (2k+1)\pi$ defines attracting limit cycles, $k = 0, 1, \ldots$.

7.   The system $x' = x(1 - 0.1x - 0.1y)$, $y' = y(2 - 0.05x - 0.025y)$ is a model for competitive interaction with overcrowding. The mass-action terms $-(x^2 + xy)/10$ and $-(xy + y^2/2)/20$ model the negative effects of large populations on the species' growth rates. See Fig. 7. It appears from the figure that the y species "wins" the competition because all orbits inside the first quadrant tend to the point $x = 0$, $y = 80$, corresponding to the extinction of the x-species. So the positive limit set of every orbit inside the population quadrant appears to be the equilibrium point $(0, 80)$.

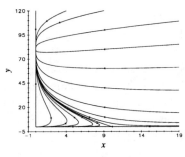

Problem 7.

9. The system is $x' = x(1 - x - 3.75y + 2xy + y^2)$, $y' = y(-1 + y + 3.75x - 2x^2 - xy)$.

    (a). The equilibrium points $(0,0)$, $(1,0)$, and $(0,1)$ appear to be saddle points, while $(1/4, 1/4)$ seems to be an unstable spiral point, $(4/3, 4/3)$ an unstable spiral point, and $(7/4, -3/4)$ a stable node. See Fig. 9(a).

    (c). See Figs. 9(c) where $c = -4$ (Graph 1) and $c = -3.5$ (Graph 2), respectively. A close observation of the orbits as they are being generated suggests that for $c < -3.75$ one of the spirals from the unstable spiral point at the upper right tends to the equilibrium point $(0,1)$ as $t \to \infty$. This means that the "saddle connection" on $x + y = 1$ (when $c = -3.75$) is broken. Similarly, when $c > -3.75$, an outward spiral from the unstable spiral point at the lower left approaches $(1,0)$, again breaking the original saddle connection.

Problem 9(a).

Problem 9(c), Graph 1.

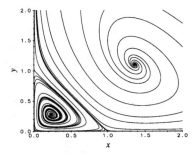

Problem 9(c), Graph 2.

11. The system is $x' = f(x, y)$, $y' = g(x, y)$, and we will scale time by $s = s(t)$ where $dt/ds = h(x, y)$ and h is a given function with a fixed sign.

    (a). Let $dt = h\, ds$. The scaled system becomes $dx/ds = fh$, $dy/ds = gh$. Since $h(x, y)$ is positive everywhere, the two systems have identical orbits and the same orientation. Since $dy/dx = g/f$ for both systems, the positive factor h affects only the rate at which orbits are generated, but not their orientation or their shape. If $h(x, y)$ is everywhere negative, the two systems have identical orbits but with opposite orientation.

    (c). The system is $x' = f(x, y) = x - 10y - x(x^2 + y^2)$, $y' = 10x + y - y(x^2 + y^2)$. Let h be the function $h = 1 - \exp[-10(x - 1)^2 - 10y^2]$. Using polar coordinates, the unscaled system becomes $r' = r - r^3$, $\theta' = 10$. This system has an attracting limit cycle $r = 1$. However, the scaled system is changed to $r' = (r - r^3)\{1 - \exp[-10(r\cos\theta - 1)^2 - 10r^2\sin^2\theta]\}$, $\theta' = 10 - 10\exp[-10(r\cos\theta - 1)^2 - 10r^2\sin^2\theta]$. We see that this scaled system has an attracting cycle graph on the unit circle. See Figs. 11(c), Graph 1 and Graph 2 for an orbit and its components of the unscaled and scaled systems, respectively. The factor $h(x, y)$ in the scaled system in x, y-coordinates is positive except at the points $(\pm 1, 0)$ on the unit circle, where h is zero. The scaled system has two new equilibrium points. All other orbits are the same in the scaled and unscaled systems. However, the component graphs of solutions of the two systems are very different; the nonlinear scale factor h has dramatic effects on the "speed" at which orbits are traced out.

Problem 11(c), Graph 1.

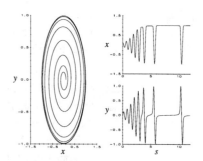
Problem 11(c), Graph 2.

13. The system is $x' = xF(x, y)$, $y' = yG(x, y)$, where $\partial F/\partial x$ and $\partial G/\partial y$ are negative in the population quadrant. Let $K = 1/xy$ for $x > 0$ and $y > 0$. Then $\partial(KxF)/\partial x + \partial(KyG)/\partial y = y^{-1}\partial F/\partial x + x^{-1}\partial G/\partial y < 0$ if $\partial F/\partial x < 0$ and $\partial G/\partial y < 0$, $x > 0$, $y > 0$. Now apply the Bendixson proof (see Background Material) to the scaled system $x' = xFK$, $y' = yGK$. The system $x' = xF$, $y' = yG$ has no cycles in the population quadrant.

Background Material: Proof of Bendixson's Negative Criterion (Theorem 9.2.4)

The system is $x' = f(x, y)$, $y' = g(x, y)$. Suppose that the quantity $\Delta = \partial f/\partial x + \partial g/\partial y$ is positive throughout a simply-connected region R. Suppose, contrary to the final assertion of Theorem 9.2.4, that a cycle or cycle-graph Γ lies inside R. We shall show that this leads to a contradiction. Applying Green's Theorem (Theorem B.5.12) to $\partial f/\partial x + \partial g/\partial y$ over the region S consisting of Γ and its interior, we have that

$$0 < \int_S\int (\partial f/\partial x + \partial g/\partial y)\, dx\, dy \quad \text{[since } \partial f/\partial x + \partial g/\partial y > 0 \text{ in } S]$$

$$= \oint_\Gamma (g\, dx - f\, dy) \quad \text{[Green's Theorem]}$$

But $dx = f\,dt$ and $dy = g\,dt$ along Γ because Γ is an orbit (if it is a cycle) or a union of orbits (if it is a cycle graph) of $x' = f$, $y' = g$. This implies that

$$g\,dx - f\,dy = gf\,dt - fg\,dt = 0$$

everywhere on Γ. We are left with the contradiction

$$0 < \oint_{\Gamma} (g\,dx - f\,dy) = 0$$

So there can be no cycle or cycle-graph with the property that it and its interior lie in the region R where the quantity Δ is positive. A similar proof applies if Δ is negative.

9.3 Bifurcations

Comments

In the last two or three decades, mathematicians, scientists, and engineers have discovered bifurcations in all sorts of dynamical systems. And it is true that bifurcation analysis often is the easiest way to handle a system where a change in a parameter leads to a sudden change in orbital or solution behavior. Nevertheless, it seems to us that it may be a case of "everything looks like a nail to a person with a hammer." With that caveat, we present in the text and in the problems some of the bifurcations most commonly observed in nonlinear systems (saddle-node, transcritical, pitchfork, Hopf). Of these, the Hopf bifurcation that spawns a limit cycle from an equilibrium point is the most interesting and useful. We present the satiable predator model, but be warned that it is not a very realistic model of population dynamics. The scroll circuit model system (see text and Problem 7) makes a nice term project for electrical engineers, who might even build and test the circuit being modeled. The scroll circuit and the model are only ten years old, and still not completely understood. Incidentally, you might also think of bifurcations in the context of elementary algebra; for example, real roots of a quadratic merging and then disappearing into the complex plane as a coefficient changes.

1. First find the equilibrium points, and then calculate the Jacobian matrix at each equilibrium point and its eigenvalues. Use the table after Section 8.2 to name the equilibrium points. We only describe the results and omit the details. In the bifurcation diagrams the solid curves indicate the x-coordinates of stable equilibria as functions of c, the dashed curves show the x-coordinates of unstable equilibria. The Jacobian calculations are omitted below. You might want to look back at Section 2.4.

(a). Saddle-node bifurcation. The system is $x' = c + 10x^2$, $y' = x - 5y$. When c is positive, there are no equilibrium points [Fig. 1(a), Graph 1]. If $c = 0$, there is a nonelementary saddle-node equilibrium point at the origin because 0 is an eigenvalue of the Jacobian matrix of the rate function of this system at the origin [Fig. 1(a), Graph 2]. As in Example 9.3.1, the nature of each equilibrium point before and after the bifurcation is determined by the eigenvalues of the Jacobian matrix at that point. According to the table at the end of Section 8.2, the system and its linearization have the same type of equilibrium point as long as the real parts of the eigenvalues of the Jacobian matrix are nonzero. From Fig. 1(a),

Graph 3, we see that as c changes from negative values through zero and into the positive range, an asymptotically stable node at $(-\sqrt{-c/10}, -\sqrt{-c/10}/5)$ and an unstable saddle at $(\sqrt{-c/10}, \sqrt{-c/10}/5)$ move toward each other, collide (when $c = 0$), and vanish. When $c = 0$, a bifurcation occurs at the saddle-node at the origin. See Figs. 1(a), Graphs 1–3 where $c = 1, 0, -1$, respectively, and Fig. 1(a), Graph 4 for the associated bifurcation diagram.

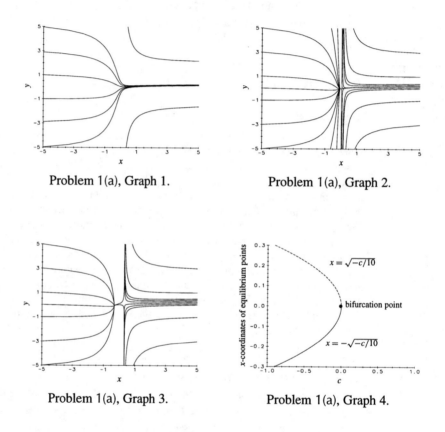

Problem 1(a), Graph 1.   Problem 1(a), Graph 2.

Problem 1(a), Graph 3.   Problem 1(a), Graph 4.

(c). Transcritical bifurcation. The system is $x' = cx + 10x^2$, $y' = x - 2y$. The system has an asymptotically stable node at the point $(-c/10, -c/20)$ and a saddle at the origin if c is positive. From Figs. 1(c), Graphs 1–3, we see that if c decreases to 0, then the two equilibrium points merge into a saddle-node at the origin. As c moves away from 0 and becomes negative, the equilibrium point $(-c/10, 0)$ and the origin exchange orbital characters, the origin becoming the asymptotically stable node and $(-c/10, -c/20)$ the saddle. A transcritical bifurcation occurs at $c = 0$. See Figs. 1(c), Graphs 1–3 where $c = -20, 0,$ and 20, respectively, and Fig. 1(c), Graph 4 for the associated transcritical bifurcation diagram.

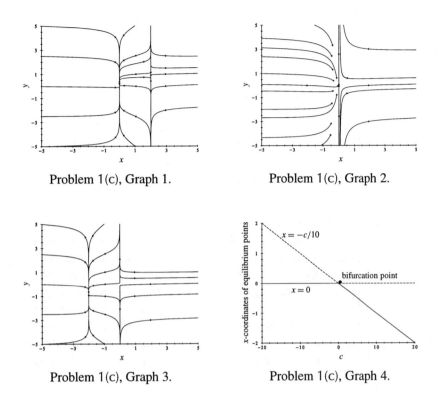

Problem 1(c), Graph 1.

Problem 1(c), Graph 2.

Problem 1(c), Graph 3.

Problem 1(c), Graph 4.

(e). Pitchfork bifurcation. The system is $x' = cx + x^5$, $y' = -y$. The system has saddles at $(\pm(-c)^{1/4}, 0)$ and an asymptotically stable node at the origin if c is negative. From Figs. 1(e), Graphs 1–3, we see that if c increases to 0, then the three equilibrium points merge into a nonlinear saddle at the origin. Also, as c moves away from 0 and becomes positive, the equilibrium points $(\pm(-c)^{1/4}, 0)$ vanish and the origin becomes a nonlinear saddle. A bifurcation occurs if $c = 0$. See Figs. 1(e), Graphs 1–3 where $c = -1, 0$, and 1, respectively, and Fig. 1(e), Graph 4 for the associated pitchfork bifurcation diagram, where (*in this graph only*) the dashed line is the x-coordinate of a *stable* equilibrium point and the solid curves the x-coordinates of *unstable* equilibrium points.

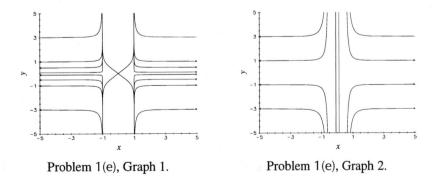

Problem 1(e), Graph 1.

Problem 1(e), Graph 2.

Problem 1(e), Graph 3.

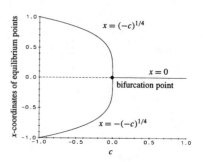

Problem 1(e), Graph 4.

3.   The system is $x' = -cx + y + x(x^2 + y^2)$, $y' = -x - cy + y(x^2 + y^2)$. The origin is the only equilibrium point. The characteristic polynomial of the Jacobian matrix of the system at the origin is $\lambda^2 + 2c\lambda + c^2 + 1$, and the eigenvalues are $\lambda = -c \pm i$. Therefore, we see that the original system bifurcates as c increases through 0 from an unstable spiral point to a stable spiral point surrounded by a repelling limit cycle. In polar coordinates, we have $r' = r(-c + r^2)$, $\theta' = -1$. So for $c > 0$ the cycle is $r = \sqrt{c}$. See Figs. 3, Graphs 1–3 where $c = -1$, 0, and 1, respectively, and Fig. 3, Graph 4 for the associated bifurcation diagram. In this graph the solid lines indicate the r-coordinate (or r-amplitude) of an *unstable* equilibrium point (or cycle); the dashed line is the r-coordinate of a *stable* equilibrium point.

Problem 3, Graph 1.

Problem 3, Graph 2.

Problem 3, Graph 3.

Problem 3, Graph 4.

5. The system is $x' = c(x - 5c) + (y - 5c) - (x - 5c)[(x - 5c)^2 + (y - 5c)^2]$, $y' = -(x - 5c) + c(y - 5c) - (y - 5c)[(x - 5c)^2 + (y - 5c)^2]$. The equilibrium point P has coordinates $x = y = 5c$.

(a). Let $u = x - 5c$, $v = y - 5c$. We obtain $u' = cu + v - u(u^2 + v^2)$ and $v' = -u + cv - v(u^2 + v^2)$, In polar form, $r' = cr - r^3$, $\theta' = -1$.

(c). See Figs. 5(c), Graphs 1–5, for $c = -0.5, 0, 0.5, 1,$ and 1.5, respectively. See also Figures 9.3.21 and 9.3.22 in the text. The asymptotically stable equilibrium point at P [for $c < 0$] moves upward and to the right as c increases. At $c = 0$, the point is still asymptotically stable [Fig. 5(c), Graph 2, and Figure 9.3.21], but there is a Hopf bifurcation as c increases beyond 0, a bifurcation to an attracting limit cycle around the destabilized P for $c > 0$. The amplitude of the limit cycle expands as c increases, while the cycle moves upward and to the right. Figure 9.3.22 shows orbits both before and after the Hopf bifurcation. The pictures in Figure 9.3.22 resemble an expanding two-dimensional smoke ring that expands and drifts upwards and to the right as c increases.

Problem 5(c), Graph 1.

Problem 5(c), Graph 2.

Problem 5(c), Graph 3.

Problem 5(c), Graph 4.

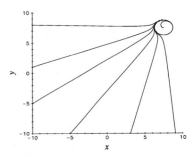

Problem 5(c), Graph 5.

7. Group project.

9.4   Chaos

Comments

Finally, the system that started it all! Lorenz's name is firmly attached to the theory of chaos, and deservedly so. It was his remarkable (but characteristically low-key) paper in 1963 that began the current multitude of studies on the implications and the very definition of chaotic dynamics. It should be noted, though, that Mary Cartwright and J. E. Littlewood did pioneering work on the subject in England in 1945. Our focus here (and in the several group projects in the problem set) is on chaotic behavior in ODEs; we have previously noted chaotic dynamics in a discrete map in Section 2.7. We prove what we can about the behavior of the Lorenz system, but ultimately must use pictures to illustrate the chaotic motion. People often have a hard time seeing where the chaos is, because the pictures seem to show fairly regular oscillations first near one equilibrium point and then near another. However, it is nearly impossible to predict just when a given orbit will be near one equilibrium point and when it will move close to the other; moreover, the number of oscillatory whirls about an equilibrium point before moving away also seems to be unpredictable. Chaos is, as noted in the section, a feature of a set of orbits, a so-called strange attractor. These strange attractors are a generalization to higher dimensions of the attracting equilibrium point, cycle, or cycle-graph limit sets of any bounded orbit of a planar autonomous system. We only give a brief mention of what strange attractors and chaotic dynamics on a strange attractor actually mean in a mathematical (rather than an informal "hand-waving") sense. The reason is that the definitions are still in a state of flux—and the students should be so informed. **Warning**: Time-state curves and orbits of the Lorenz system, the Rössler system (Problem 3), the Duffing equation (Problem 4), and the driven and damped pendulum (Problem 5) are sensitive to changes in initial data. Since numerically computed solutions are never exact, this means that these solutions may be reasonably accurate only over a short time span. However, the orbits and solutions computed in this section are thought to be near a strange attractor that attracts all nearby solutions. So the computed solutions, orbits, and Poincaré time sections shown here are each most likely composites of several orbits or time-state curves that (at least for the end segment of the solution interval) are essentially in the strange attractor. Although we may not be looking at the actual orbit or time-state curve through the initial point, we do get a pretty good view of the strange attractor.

---

1. **(a).** The equilibrium points P_1 and P_2 are easily found by equating the right sides of the equation of the Lorenz system (1) to zero and solving for x, y, and z.

(c). At P_1 (and P_2), the eigenvalues of the Jacobian matrix are the roots of $\lambda^3 + (1 + b + \sigma)\lambda^2 + (\sigma + r)b\lambda + 2\sigma b(r - 1)$. According to the Routh Test (Theorem 7.9.7) all roots of this cubic have negative real parts if and only if the coefficients are positive (which they are because b and σ are positive and $r > 1$) *and* if $(1 + b + \sigma)(\sigma + r)b > 2\sigma b(r - 1)$. This equality reduces to $\sigma(\sigma + b + 3) > r(\sigma - b - 1)$, i.e., $r < r_c = \sigma(\sigma + b + 3)/(\sigma - b - 1)$ if $\sigma > b + 1$. P_1 and P_2 are locally asymptotically stable if $1 < r < r_c$ since all the eigenvalues have negative real parts for these values of r, while if $r > r_c$ both P_1 and P_2 are unstable.

3. Group project.

5. Group project.

Background Material: Verification of the Lorenz Squeeze

To show the squeeze property we argue informally along the following lines. Suppose S is a bounded region of state space with positive volume $V(S)$ and bounding surface ∂S, and suppose that ∂S is smooth, connected, and closed. For example, S could be a solid ball with spherical surface ∂S. Let ∂S_t denote the bounding surface of S_t, and let $V(S_t)$ denote the volume of the region S_t. As t increases from 0, S_t and ∂S_t will look less and less like $S = S_0$ and $\partial S = \partial S_0$. Interpreting the Lorenz equations as the equations of motion of the particles of a gas or a fluid, we have that

$$\frac{dV(S_t)}{dt} = \left\{ \begin{array}{c} \text{net volume of fluid} \\ \text{crossing } S_t \text{ in the} \\ \text{unit outward normal} \\ \text{direction per unit of} \\ \text{time} \end{array} \right\} = \iint_{\partial S_t} \mathbf{F} \cdot \mathbf{n}\, dA \qquad \text{(i)}$$

where $\mathbf{F}$ is the vector field defined by the right-hand side of the Lorenz system [text system (1)], $\mathbf{n}$ is a unit outward normal vector field to the bounding surface ∂S_t, and the integral is a surface integral. By the Divergence Theorem of vector calculus (Theorem B.5.13), the surface integral becomes the volume integral of the divergence over S_t:

$$\iint_{\partial S_t} \mathbf{F} \cdot \mathbf{n}\, dA = \iiint_{S_t} \nabla \cdot \mathbf{F}\, dV \qquad \text{(ii)}$$

The divergence of $\mathbf{F}$ is given by

$$\nabla \cdot \mathbf{F} = \frac{\partial F_1}{\partial x} + \frac{\partial F_2}{\partial y} + \frac{\partial F_3}{\partial x}$$

where F_1, F_2, F_3 are given by the right-hand sides of the equations of text system (1). We have that $\nabla \cdot \mathbf{F} = -\sigma - 1 - b$. From (i) and (ii) we have that

$$\frac{dV(S_t)}{dt} = \iiint_{S_t} -(\sigma + 1 + b)\, dV = -(\sigma + 1 + b)V(S_t) \qquad \text{(iii)}$$

Solving the scalar ODE $V' = -(\sigma + 1 + b)V$, with initial condition $V(0) = V(S)$ when $t = 0$, we have that

$$V(S_t) = V(S)e^{-(\sigma + 1 + b)t} \qquad \text{(iv)}$$

We see that as $t \to \infty$, $V(S_t) \to 0$.

Fourier Series and Separation of Variables for PDEs

This chapter begins with the derivation of a boundary/initial value problem for the PDE which models a vibrating guitar string. Later in the chapter we develop a method for constructing a solution formula for that problem. In between we lay the groundwork for deriving solution formulas for a class of linear boundary/initial value problems. The intervening sections contain material on Fourier series and Sturm-Liouville problems that is needed to solve the boundary/initial value problems of this chapter.

Section 10.1 was written to be as self-contained as possible in order to provide a brief introduction and example of the Method of Separation of Variables. One could quit at this point and still come away with a good idea of what this method is all about. Sections 10.2 and 10.3 set the stage for Fourier Series in order to give a clean introduction to Fourier Trigonometrical Series in Sections 10.4 and 10.5. Sturm-Liouville Problems are treated in a unified way in Section 10.6. It is a short section and should be read before reading Section 10.7 on eigenfunction expansions and the Method of Separation of Variables. In Section 10.8 we discuss a model PDE for heat conduction, and apply the eigenfunction techniques for the heat equation to discuss the changing temperatures in a rod and in an underground storage cellar. In Section 10.9 we consider Laplace's PDE as a model for steady-state temperatures in a body, given the temperatures at all the boundary points. Fourier series, eigenfunction expansions, and Separation of Variables underlie all we do in this chapter.

10.1 Vibrations of a String

Comments

We use the PDE of a vibrating string as the introduction to PDEs, separation of variables, and Fourier series because the twanging of a guitar string is familiar to everyone. Our approach is to start with a derivation of the wave equation for the transverse motion of the string, introduce the three different kinds of linear boundary conditions and the corresponding boundary operators, and then focus on separating the PDE to get the standing waves modulated by sinusoidal functions of time. We work out the details of a boundary problem that involves fixed endpoints (e.g., the guitar string fastened at each end). We show how the coefficients are calculated by the Fourier method, but dodge all questions of convergence.

Everything is straightforward, but the problems tend to be somewhat lengthy. If you don't have access to a CAS, use integral tables or other sources to calculate the integrals. You need to develop early on a standard format for doing a separation of variables construction. Otherwise you will lose your way in the lengthy process. Be sure to look at the steps leading up to (10) where the variables are separated, the separation constant λ is introduced, and separate ODEs are found (one with boundary conditions).

---

1. The model equations for a vibrating string with free boundary conditions are based on system (8) in the text and the condition given in (6) [with $\gamma = 0$]. We use (10) for the separated functions $X(x)$ and $T(t)$. Putting all this together, we have the model:

$$u_{tt} - c^2 u_{xx} = 0, \qquad 0 < x < L, \quad t > 0$$
$$u_x(0, t) = 0, \qquad t \geq 0$$
$$u_x(L, t) = 0, \qquad t \geq 0$$
$$u(x, 0) = f(x), \qquad 0 \leq x \leq L$$
$$u_t(x, 0) = g(x), \qquad 0 \leq x \leq L$$

(a). The standing waves $X(x)T(t)$ must satisfy the pair of equations $X'' = \lambda X$, $T'' = c^2 \lambda T$, for a constant λ. We ignore the initial conditions for $u(x, 0)$, $u_t(x, 0)$ throughout. For the standing waves to satisfy the free boundary conditions $u_x(0, t) = 0$, $u_x(L, t) = 0$, for all $t \geq 0$, we must have $X'(0) = X'(L) = 0$. Moreover, if $X(x)$ satisfies the equation $X'' = \lambda X$, for some constant λ, and the endpoint conditions $X'(0) = X'(L) = 0$, then integration by parts [see (13)] yields that $\int_0^L X(x) X''(x) dx = -\int_0^L (X'(x))^2 dx$. But since $\int_0^L X(x) X''(x) dx = \lambda \int_0^L (X(x))^2 dx$, we see that $\lambda \leq 0$. Now if $\lambda = 0$, then $X'' = 0$, so $X(x)$ has the form $Ax + B$ for some constants A, B, not both zero. To satisfy the conditions $X'(0) = X'(L) = 0$ we must take $A = 0$. So $X_0(x) = 1$, $0 \leq x \leq L$, is one of the functions that models the space dependence of a standing wave with free endpoint conditions. Similarly, if $\lambda = -k^2$, $k \neq 0$, then $X(x)$ must have the form $A \sin kx + B \cos kx$, for constants A, B, not both zero. To satisfy the free boundary conditions at $x = 0$ and $x = L$ we see that $k(A \cos kx - B \sin kx) = 0$ at $x = 0$ and $x = L$. At $x = 0$, we must have $kA = 0$, so $A = 0$, while at $x = L$, we then must have $-kB \sin kL = 0$. Since we assume that $k \neq 0$ here and $B = 0$ would give only the trivial solution, we see that $\sin kL = 0$. This implies that $kL = n\pi$, for $n = 1, 2, \ldots$, providing the further candidates for the space dependence of standing waves, $X_n = \cos n\pi x/L$, $n = 1, 2, \ldots$. So we have that the standing waves are

$$u_n(x, t) = \cos \frac{n\pi x}{L} \left(A_n \cos \frac{n\pi ct}{L} + B_n \sin \frac{n\pi ct}{L} \right), \qquad n = 0, 1, 2, \ldots$$

where A_n, B_n are arbitrary constants. Note that if $n = 0$, $u_0 = X_0(x)T_0(t) = A_0 + B_0 t$ since $X_0 = 1$ (see above) and $T''(t) = 0$. This gives all of the standing waves with free boundary conditions at the endpoints $x = 0$ and $x = L$.

3. Suppose that u and v both solve the boundary/initial problem

$$\text{(PDE)} \quad u_{tt} = c^2 u_{xx} + F(x, t), \qquad\qquad 0 < x < L, \quad t > 0$$

$$\text{(BC)} \quad u(0, t) = \alpha(t), \qquad u(L, t) = \beta(t), \quad t \geq 0$$

$$\text{(IC)} \quad u(x, 0) = \gamma(x), \qquad u_t(x, 0) = \delta(x), \quad 0 \leq x \leq L$$

We want to show $u(x, t) = v(x, t)$. Note that $U(x, t) = u(x, t) - v(x, t)$ solves the problem

$$\begin{cases} U_{tt} = c^2 U_{xx}, & 0 < x < L, \quad t > 0 \\ U(0, t) = U(L, t) = 0, & t \geq 0 \\ U(x, 0) = U_t(x, 0) = 0, & 0 \leq x \leq L \end{cases}$$

Defining the function

$$W(t) = \frac{1}{2} \int_0^L (U_t^2 + c^2 U_x^2) \, dx$$

we see that

$$W'(t) = \frac{1}{2} \int_0^L \frac{d}{dt} \left(U_t^2 + c^2 U_x^2 \right) dx = \int_0^L (U_t U_{tt} + c^2 U_x U_{xt}) \, dx$$

$$= \int_0^L [U_t(c^2 U_{xx}) + c^2((U_x U_t)_x - U_{xx} U_t)] \, dx$$

$$= c^2 \int_0^L (U_x U_t)_x \, dx = U_x(L, t) U_t(L, t) - U_x(0, t) U_t(0, t)$$

Now since $U(0, t) = 0$ and $U(L, t) = 0$, $t \geq 0$, we can compute $U_t(0, t)$ and $U_t(L, t)$ by just differentiating $U(0, t)$ and $U(L, t)$ with respect to t (from the definition of the partial derivative). So $U_t(0, t) = U_t(L, t) = 0$ and $W'(t) = 0$, for all $t \geq 0$; the integrand in the definition of W must vanish for all (x, t) in R (since the integrand is nonnegative), where R is the region $0 < x < L$, $t > 0$. From the Mean Value Theorem (Theorem B.5.4) we conclude that U is a constant on the region, $0 \leq x \leq L$, $t \geq 0$. Evaluating this constant on the boundary of R where $u = v$, we conclude that $U = 0$ on R, and so $u = v$ on R, as was asserted.

10.2 Orthogonal Functions

Comments

A better title for this section might be "Everything You Wanted to Know about Scalar Product Spaces with Orthogonal Bases." The modern approach to Fourier Series is based on the material of this section. Without it, Fourier Series reduces to endless "turn-the-crank" computations of Fourier-Euler coefficients, and there is little chance of understanding the "how" and the "why" of it all. In the few days you have available in the course to do "all" of Fourier series, PDEs, and separation of variables you won't have the time (or the inclination) to be very mathematical. So how to solve this dilemma? Here is what you might do: focus on examples, skip theory, work problems, focus on the close relationship of all of this material to standard Euclidean concepts (e.g., the Parseval Relation (10) is just an extension of the Pythagorean theorem that the square on the hypotenuse is the sum of the squares on the sides). No harm is done by this process of "nonrigorizing" everything in your introductory studies of Fourier Analysis. In fact, as a first approach to some deep ideas, this low-level, low-key approach is the best way to do it. So don't hesitate to focus on the ideas and the examples, and not too much on the theory.

---

1. Using the identity $\sin \alpha \sin \beta = [\cos(\alpha - \beta) - \cos(\alpha + \beta)]/2$, we see that

$$\int_0^L \sin \frac{n\pi x}{L} \sin \frac{m\pi x}{L} = \frac{1}{2} \int_0^L \cos \frac{(n-m)\pi x}{L} dx - \frac{1}{2} \int_0^L \cos \frac{(n+m)\pi x}{L} dx$$

If $m = n = 0$, the integrand and the integral are zero. Now, if $n \neq m$, then the right-hand side becomes

$$\frac{1}{2} \frac{L}{(n-m)\pi} \sin \frac{(n-m)\pi x}{L} \Big|_0^L - \frac{1}{2} \frac{L}{(n+m)\pi} \sin \frac{(n+m)\pi x}{L} \Big|_0^L = 0$$

On the other hand, if $n = m \neq 0$, then the right-hand side of the identity becomes

$$\frac{1}{2} \int_0^L dx - \frac{L}{4n\pi} \sin \frac{2n\pi x}{L} \Big|_0^L = \frac{L}{2}$$

The second set of formulas in (4) is proved in the same way, using the identity $\cos \alpha \cos \beta = [\cos(\alpha - \beta) + \cos(\alpha + \beta)]/2$.

3. The plucked guitar string problem in this case is

$$\begin{aligned}
&u_{tt} = c^2 u_{xx} & &0 < x < \pi, \quad t > 0, \quad c^2 = T/\rho \\
&u(0, t) = 0, \quad u(\pi, t) = 0, & &t \geq 0 \\
&u(x, 0) = f(x), \quad u_t(x, 0) = 0, & &0 \leq x \leq \pi
\end{aligned}$$

According to text formulas (5) and (6), the solution is

$$u(x, t) = \sum_{n=1}^{\infty} A_n \sin nx \cos nct$$

$$A_n = \frac{2}{\pi} \int_0^\pi f(x) \sin nx \, dx, \qquad n = 1, 2, \ldots$$

So to solve parts (a), and (c), all we need to do is calculate A_n, $n = 1, 2, \ldots$, for the particular $f(x)$ given.

(a). If $f(x) = \sin x$, then $A_1 = (2/\pi) \int_0^\pi \sin^2 x \, dx = 1$. $A_n = 0$ for $n > 1$ because by text formula (4) $\sin x$ and $\sin nx$, $n > 1$, are orthogonal. So $u(x, t) = \sin x \cos ct$, as one would expect.

(c). Let $f(x) = \sum_{n=1}^{10} (\sin nx)/n$. Then for $n = 1, \ldots, 10$, $A_n = (2/\pi) \int_0^\pi [\sin(nx)^2/n] dx$ [by (4)]. Also by (4), $A_n = 0$ for $n = 11, 12, \ldots$. So $u(x, t) = \sum_{n=1}^{10} (\sin nx \cos nct)/n$.

5. Define $\langle f, g \rangle = \int_0^1 f(x)g(x)e^x \, dx$.

(a). We want to show that $\langle f, g \rangle$ is a scalar product. Since $f(x)g(x) = g(x)f(x)$ for all scalar functions f and g, and since the functions in $\mathbf{C}^0[0, 1]$ are real, we have that

$$\int_0^1 f(x)g(x)e^x\, dx = \int_0^1 g(x)f(x)e^x\, dx$$

and $\langle f, g \rangle = \langle g, f \rangle$, so the first condition (8a) of a scalar product is met. The property of bilinearity [(8b)], $\langle \alpha f + \beta g, h \rangle = \alpha \langle f, h \rangle + \beta \langle g, h \rangle$ follows immediately from the corresponding property for the integral of a product. To show positive definiteness [(8c)], observe that $\langle f, f \rangle = \int_0^1 f^2(x)e^x\, dx$. If $f(x) = 0$ for all x in $[0, 1]$, then $\langle f, f \rangle = 0$. Integrals of nonnegative continuous functions satisfy the integral mean value theorem (see Theorem B.5.4), so for some x_0 in the interval $[0, 1]$, $\langle f, f \rangle = f^2(x_0)e^{x_0}$, which is certainly nonnegative. Suppose that $\langle f, f \rangle = \int_0^1 f^2(x)e^x\, dx = 0$, but that $f(x_0) \neq 0$ for some x_0, say $f(x_0) > 0$. Then there is an interval $I = [c, d]$ (maybe quite small) around x_0 and $f(x) > 0$ for all x in $[c, d]$. But then $\int_c^d f^2(x)e^x\, dx = f^2(x_1)e^{x_1}(d - c) > 0$ for some x_1 in $[c, d]$. So $\int_0^1 = \int_0^c + \int_c^d + \int_d^1 > 0$; the supposition that $f(x_0) > 0$ for some x_0 is wrong. So $\langle f, g \rangle$ meets all three conditions for an inner product.

(c). Tables of integrals are used to compute the integrals, but a CAS could be used as well.

$$\langle x, 1 - x \rangle = \int_0^1 x(1 - x)e^x\, dx = 3 - e$$

$$\langle e^{-x/2}\sin(\pi x/2), e^{-x/2}\sin(3\pi x/2) \rangle = \int_0^1 \sin(\pi x/2)\sin(3\pi x/2)\, dx = 0$$

$$\langle \cos(\pi x/2), 1 \rangle = \int_0^1 \cos(\pi x/2)e^x\, dx = (2\pi e - 4)/(\pi^2 + 4)$$

7. **(a).** Using the properties of the norm and the scalar product given in the text, we have that $\|u + v\|^2 + \|u - v\|^2 = \langle u + v, u + v \rangle + \langle u - v, u - v \rangle = \langle u, u \rangle + \langle u, v \rangle + \langle v, u \rangle + \langle v, v \rangle + \langle u, u \rangle - \langle u, v \rangle - \langle v, u \rangle + \langle v, v \rangle = 2\langle u, u \rangle + 2\langle v, v \rangle = 2\|u\|^2 + 2\|v\|^2$, which was to be shown.

(c). The proof that $\langle u, v \rangle = (\|u + v\|^2 - \|u - v\|^2)/4$ follows by direct computation of the right-hand side.

9. **(a).** For u in E, consider the vectors $v = \text{proj}_S(u)$ and $w = u - \text{proj}_S(u)$. Then $v + w = u$ and v is in S, where S is the span of the orthogonal set $\{u^1, \ldots, u^n\}$. Since

$$\langle u^k, w \rangle = \langle u^k, u - \sum_1^n [\langle u, u^i \rangle / \|u^i\|^2]u^i \rangle = \langle u^k, u \rangle - [\langle u, u^k \rangle / \|u^k\|^2]\langle u^k, u^k \rangle = 0$$

for each $k = 1, 2, \ldots, n$, it follows that w is orthogonal to S, and the assertion follows. The proof of uniqueness is omitted.

(c). We want to find the distance from $(1, 1, 1)$ to the plane $x + 2y - z = 0$. The plane $x + 2y - z = 0$ contains the two vectors $u = (1, 0, 1)$ and $v = (0, 1, 2)$. Now if v is regarded as a basis for the line S through the origin parallel to v then $w = u - \text{proj}_S(u)$ also lies in the plane and the pair $\{v, w\}$ is a basis for the plane P. We have that $w = (1, 0, 1) -$

$2(0, 1, 2)/5 = (1, -2/5, 1/5)$. To find the distance d between the point $q = (1, 1, 1)$ and the plane P we see from the statement of part **(b)** that this distance is given by $d = \|q - \text{proj}_S(q)\|$. Since

$$\text{proj}_S(q) = \frac{\langle q, v \rangle}{\|v\|^2} v + \frac{\langle q, w \rangle}{\|w\|^2} w = (1/3)(2, 1, 4)$$

and so $q - \text{proj}_S(q) = \frac{1}{3}(1, 2, -1)$, giving the result that $d = \|\frac{1}{3}(1, 2, -1)\| = \sqrt{6}/3$.

Background Material: Linear Spaces

$\mathbb{R}^3$ is an example of a linear space, a collection of objects, called vectors, for which vector addition and scalar multiplication are defined and satisfy the properties:

❖ **Vector Space Properties.** In the nonempty collection V of objects (called *vectors*) there are two operations defined which satisfy the properties listed below. Denote by "+" the operation of *addition* which assigns a vector $u + v$ to every pair of vectors u and v. For a given set of *scalars*, the operation of *scalar multiplication* assigns to every scalar α and every vector u the vector αu. With these operations V has the properties:

- (A1) *Addition is order independent.* A finite collection of vectors can be added together in any order without affecting the sum.

- (A2) *Existence of zero element.*[1] There is an additive *zero* element in V, denoted by 0, with the property that $u + 0 = u$ for all vectors u.

- (A3) *Existence of additive inverses.* Every vector u has an *additive inverse*, denoted by $-u$, with the property that $u + (-u) = 0$.

- (M1) *Scalar multiplication is distributive.* For any vectors u and v and any scalars α and β, we have that $\alpha(u + v) = \alpha u + \alpha v$ and $(\alpha + \beta)u = \alpha u + \beta u$.

- (M2) *Scalar multiplication is associative.* For any vectors u and any scalars α and β, we have that $\alpha(\beta u) = (\alpha\beta)u$.

- (M3) For the scalar 1 we have that $1u = u$, for all vectors u.

The Vector Space Properties define the abstract notion of a linear (or vector) space.

❖ **Linear Space.** Let V be a set of objects and **F** a set of scalars (**F** is either **R**, the real numbers, or **C**, the complex numbers) with two operations, "addition" and "scalar multiplication," defined in V such that the Vector Space Properties hold. Then V is a *real linear* (or *vector*) *space* if **F** = **R**, and a *complex linear* (or *vector*) *space* if **F** = **C**. The elements of V are sometimes called *vectors*.

---

[1]It is traditional to use the same symbol 0 for the zero scalar and the zero vector—context distinguishes one from the other.

A feature of this abstract approach to linear spaces is that we can deduce from the defining properties (A1)–(A3) and (M1)–(M3) some additional properties which must then hold in *any* linear space. Some of them are as follows:

THEOREM

Further Properties of Linear Spaces. Let V be any linear space.

1. There is only one zero vector in V.

2. The vector αu is zero if either the scalar $\alpha = 0$ or the vector $u = 0$. The other way around, if $\alpha u = 0$, then either $\alpha = 0$ or $u = 0$.

3. Every vector v in V has a *unique* additive inverse, $-v$; moreover $-v = (-1)v$ for any vector v.

Each assertion is verified below:

1. Suppose that 0 and $0'$ are two zero vectors for V; then by (A1) and (A2) we see that $0 = 0 + 0' = 0' + 0 = 0'$, so there is only one zero in V.

2. For any u note that $u = (1 + 0)u = u + 0u$ [using (M1) and (M3)]. Adding $-u$ to both sides of this equation, we obtain by (A3) that $0 = 0u$, as asserted. Next, we assume that $\alpha \neq 0$. Then for any vector u, $\alpha 0 + u = \alpha 0 + (\alpha \cdot \alpha^{-1})u = \alpha(0 + \alpha^{-1}u) = \alpha(\alpha^{-1}u) = 1u = u$, where we have used (M1)–(M3). So $\alpha 0$ acts like a zero vector, and since there is only one such vector, we must have $\alpha 0 = 0$. Finally, suppose that $\alpha u = 0$ for some scalar α and some vector u. If $\alpha \neq 0$, then $u = \alpha^{-1}(\alpha u) = \alpha^{-1}0 = 0$.

3. Suppose that v and w are two additive inverses for the vector u. Then $v = v + 0 = v + (u + w) = (v + u) + w = 0 + w = w$ where we have used (A1) and (A2). So u has a unique inverse. Now since $0 = 0u = (1 - 1)u = 1u + (-1)u = u + (-1)u$, we see that $-u = (-1)u$, finishing the proof.

It is worth noting that there is a linear space V with only one element. In view of the result above, we see that that single element must be the zero vector; V is written as $\{0\}$ and is referred to as the *trivial* linear space. Here are two familiar examples.

EXAMPLE

Real n-tuple Space
Let $\mathbb{R}^n$ for any $n = 1, 2, \ldots$ be the collection of all ordered n-tuples of real numbers, $x = (x_1, x_2, \ldots, x_n)$. Note that x here denotes the n-tuple and the x_i are real numbers. Two elements x and y in $\mathbb{R}^n$ are identified if and only if $x_i = y_i$, $i = 1, 2, \ldots, n$. Using the real numbers $\mathbb{R}$ as scalars and defining the operations of addition and multiplication by scalars component-wise by

$$x + y = (x_1 + y_1, \ldots, x_n + y_n), \qquad \alpha x = (\alpha x_1, \ldots, \alpha x_n) \tag{i}$$

for all x, y in $\mathbb{R}^n$ and all scalars α, it is straightforward to verify that $\mathbb{R}^n$ is a real linear space. Note that $\mathbb{R}^1$ can be identified with $\mathbb{R}$, so $\mathbb{R}$ itself can be considered a real linear

| space.

EXAMPLE

Complex n-tuple Space

Let $\mathbf{C}^n$ for any $n = 1, 2, \ldots$ be the collection of all ordered n-tuples of complex numbers $z = (z_1, z_2, \ldots, z_n)$. Using the complex numbers $\mathbf{C}$ as scalars and defining operations of addition and multiplication by scalars as in (i), it is straightforward to show that $\mathbf{C}^n$ is a complex linear space. Note that $\mathbf{C}^1$ can be identified with $\mathbf{C}$.

10.3 Fourier Series and Mean Approximation

Comments

This section contains the basic ideas about the Fourier expansion of a function with respect to an orthogonal set. We start with a Euclidean space and an orthogonal set in it. We show how to construct the Fourier series of a function in the Euclidean space with respect to the orthogonal set, explain mean convergence and apply that idea to the convergence of a Fourier series of a function to the function itself. The idea of a basis comes in along the way. We explain what is meant when it is said that "a partial sum of the Fourier series of a function is the best approximation in the mean to the function."

1. This problem is solved by using Theorem 10.3.1 with $\Phi_N = \{1, \cos x, \sin x, \ldots, \cos Nx, \sin Nx\}$. The element which is closest to $f(x) = x$, $-\pi < x < \pi$, in the subspace S_N of $PC[-\pi, \pi]$ is given by

$$\text{proj}_{S_N}(f) = a_0 + \sum_{k=1}^{N}(a_k \cos kx + b_k \sin kx)$$

$$a_k = \frac{\langle f, \cos kx \rangle}{\| \cos kx \|^2} = \begin{cases} (2\pi)^{-1} \int_{-\pi}^{\pi} x \, dx, & \text{for } k = 0 \\ (\pi)^{-1} \int_{-\pi}^{\pi} x \cos kx \, dx, & \text{for } k = 1, 2, \ldots, N \end{cases}$$

$$b_k = \frac{\langle f, \sin kx \rangle}{\| \sin kx \|^2} = \frac{1}{\pi} \int_{-\pi}^{\pi} x \sin kx \, dx, \quad \text{for } k = 1, 2, \ldots, N$$

Notice that f is odd about the origin, so $f(x) \cos kx$ is also odd about the origin for all $k = 0, 1, 2, \ldots, N$. We have $a_k = 0$ for $k = 0, 1, 2, \ldots, N$. Note that $f(x) \sin kx$ is even about the origin, so

$$b_k = \frac{2}{\pi} \int_0^{\pi} x \sin kx \, dx = -\frac{2}{k\pi} \int_0^{\pi} x(\cos kx)' dx$$

$$= -\frac{2}{k\pi} \left[x \cos kx \Big|_0^{\pi} - \int_0^{\pi} \cos kx \, dx \right] = \frac{(-1)^{k+1} 2}{k}, \quad \text{for } k = 1, 2, \ldots, N,$$

So $\text{proj}_{S_n}(f) = 2 \sum_{k=1}^{N}(-1)^{k+1} k^{-1} \sin kx$ is the element in S_N closest to $f(x) = x$.

3. Here the orthogonal set is $\Phi = \{1, \cos x, \sin x \ldots, \cos nx, \sin nx, \ldots\}$ in the Euclidean space $PC[-\pi, \pi]$, and we want to find the Fourier series of f with respect to Φ for the three engineering functions given.

(a). The sawtooth function $f(x)$ is an odd function, so FS[f] contains only sine terms. Using the fact that $f(x) \sin kx$ is an even function for all k, we see that

$$B_k = \frac{2}{\pi} \int_0^\pi A(1 - \frac{x}{\pi}) \sin kx dx = \frac{-2A}{k\pi} \int_0^\pi (1 - \frac{x}{\pi})(\cos kx)' dx$$

$$= \frac{-2A}{k\pi} \left\{ (1 - \frac{x}{\pi}) \cos kx \Big|_0^\pi - \int_0^\pi \left(-\frac{1}{\pi}\right) \cos kx dx \right\} = \frac{2A}{k\pi}$$

We have $\text{FS}_\Phi[f] = (2A/\pi) \sum_{k=1}^\infty (\sin kx)/k$.

(c). The inverted triangular function $f(x)$ is even about the origin, so FS[f] does not contain any sine terms.

$$A_0 = \frac{2}{2\pi} \int_0^\pi x dx = \frac{1}{2\pi} x^2 \Big|_0^\pi = \frac{\pi}{2}$$

$$A_k = \frac{2}{\pi} \int_0^\pi x \cos kx dx = \frac{2}{k\pi} \int_0^\pi x(\sin kx)' dx$$

$$= \frac{2}{k\pi} \left\{ x \sin kx \Big|_0^\pi - \int_0^\pi \sin kx dx \right\}$$

$$= -\frac{2}{k^2\pi}(1 - (-1)^k) = \begin{cases} 0, & \text{even } k \\ -4/k^2\pi, & \text{odd } k \end{cases}$$

We have that

$$\text{FS}_\Phi[f] = \frac{\pi}{2} - \frac{4}{\pi} \sum_{m=0}^\infty \frac{\cos(2m+1)x}{(2m+1)^2}$$

5. We want to show that two functions f and g with the same Fourier series with respect to an orthogonal basis Φ of a Euclidean space are "equal" in that space, where by equal we mean that $\|f - g\| = 0$. Let $w = f - g$. Since $\Phi = \{\phi_1, \phi_2, \ldots\}$ is a basis for E, the Fourier series of w over Φ must converge to w in the mean; that is,

$$w_N = \sum_{k=1}^N \frac{\langle w, \phi_k \rangle}{\|\phi_k\|^2} \phi_k \to w \text{ as } N \to \infty$$

By the continuity of the scalar product, $\langle w_N, w \rangle \to \|w\|^2$, as $N \to \infty$. But since $\langle w, \phi_k \rangle = 0$ for all $k = 1, 2, \ldots$, we see that $w_N = 0$ for all N, so $\langle w_N, w \rangle \to 0$ as $N \to \infty$. But $\langle w_N, w \rangle \to \langle w, w \rangle$ by Theorem 10.3.3, so $\langle w, w \rangle = \|w\|^2$, so $\|w\|^2 = 0$, which in turn implies that $f = g$.

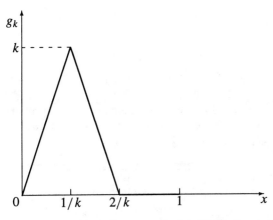

Graph 1. Sequence $\{g_k\}$ in PC[0,1].

Graph 2. Sequence $\{h_k\}$ in PC[0,1], where $k = 2^n + m$.

Background Material: Mean versus Pointwise Convergence

Mean Convergence versus Pointwise Convergence

Mean convergence of a sequence $\{f_n(x)\}$ in PC[a, b] implies the existence of an element f in PC[a, b] such that $\int_a^b (f_n(x) - f(x))^2\, dx \to 0$ as $n \to \infty$. Consider the sequence g_k in PC[0, 1], $k = 1, 2, \ldots$ defined in Graph 1. We claim that $\{g_k\}$ does *not* converge in the mean. Indeed, say $g_k \to g$, for some g in PC[0,1]. Writing $g_k = (g_k - g) + g$ and applying the Triangle Inequality, we have $\|g_k\| \le \|g_k - g\| + \|g\|$, so the sequence $\{\|g_k\|\}$ is bounded. But since $\|g_k\|^2 = \int_0^1 |g_k|^2\, dx = 2k/3 \to \infty$, as $k \to \infty$, it follows that $\{\|g_k\|\}$ is unbounded, so $\{g_k\}$ cannot converge.

On the other hand, consider the sequence h_k in PC[0,1], $k = 1, 2, \ldots$ defined in Graph 2. Note that h_k, for $k = 2^n + m$, has the value 1 inside the interval $[(m - 1)/2^n, m/2^n]$ and vanishes otherwise in [0,1]. So $\int_0^1 |h_{2^n+m}|\, dx = 1/2^n$ for all $n = 1, 2, \ldots$, and all $m = 1, 2, \ldots, 2^n - 1$. It follows that $\|h_k\|^2 = \|h_k - 0\|^2 \to 0$ as $k \to \infty$. So $\{h_k\}$ converges in the mean to the zero function of [0,1], even though for each k there is an interval on which h_k has the value one.

Pointwise, Uniform, and Mean Convergence

The sequence $\{g_k\}$ above converges pointwise to the zero function on [0,1]. On the other hand, since $g_k(1/k) = k$ it is clear that the graphs of the $g_k(x)$ cannot eventually all lie in every tube about the zero function on [0,1]. So $\{g_k\}$ does *not* converge uniformly on [0,1]. It is true, however, that $\{g_k\}$ converges uniformly to the zero function on any interval [δ, 1], where $0 < \delta < 1$. The second sequence $\{h_k\}$ considered above does not converge pointwise for *any* point in [0,1]. Indeed, for any x_0 in [0,1] and any positive integer K, there are integers $r > K$, $s > K$ such that $h_r(x_0) = 0$ and $h_s(x_0) = 1$. So $\{h_k(x_0)\}$ cannot converge at all.

10.4 Fourier Trigonometric Series

Comments

The basic theorems of "classical" Fourier Trigonometric Series are introduced in this section and in its problem set for the orthogonal set $\Phi = \{1, \cos\frac{\pi x}{T}, \sin\frac{\pi x}{T}, \ldots, \cos\frac{n\pi x}{T}, \sin\frac{n\pi x}{T}, \ldots\}$ in the Euclidean space $\text{PC}[-T, T]$. Pointwise convergence, uniform convergence, decay estimates—all of this is here along with the verification. This section and the previous two form a unit that can be used for self-study of the basics of the mathematical approach to Fourier Series.

---

1. Each of these problems can be solved without having to integrate to find Fourier coefficients.

    (a). The rearrangement $5 - \sin x - 7\sin 3x - 4\cos 6x$ already has the form of a Fourier Series (2), with $T = \pi$. So $A_0 = 5$, $B_1 = -1$, $B_3 = -7$, $A_6 = -4$, and all other coefficients are zero.

    (c). Using trigonometric identities in Appendix B.4, we see that $\sin^3 x = \sin x(1 - \cos 2x)/2 = (\sin x - \sin x\cos 2x)/2 = (\sin x)/2 - (\sin 3x - \sin x)/4 = (3\sin x)/4 - (\sin 3x)/4$ which has the form of text formula (2) with $T = \pi$. So $B_1 = 3/4$, $B_3 = -1/4$, and all other Fourier coefficients are zero.

3. We show that the function $\sin x$ has very different Fourier Series, depending on which interval $[a, b]$ we use.

    (a). The function $f(x) = \sin x$, $-\pi \leq x \leq \pi$ has the Fourier Series $\text{FS}[f] = \sin x$.

    (c). The function $f(x) = \sin x$, $-3\pi/2 \leq x \leq 3\pi/2$, is odd about the origin. By (2), (3) with $T = 3\pi/2$, $\text{FS}[f]$ has the form $\text{FS}[f] = \sum_{k=1}^{\infty} B_k \sin(2kx/3)$, where

$$B_k = \frac{2}{3\pi}\int_{-3\pi/2}^{3\pi/2} \sin x \sin\frac{2kx}{3}dx = \frac{2}{3\pi}\int_0^{3\pi/2}\left\{\cos(\frac{2k}{3} - 1)x - \cos(\frac{2k}{3} + 1)x)\right\}dx$$

$$= \frac{2}{3\pi}\left\{\frac{\sin(k\pi - 3\pi/2)}{2k/3 - 1} - \frac{\sin(k\pi + 3\pi/2)}{2k/3 + 1}\right\} = \frac{8(-1)^k k}{\pi(4k^2 - 9)}$$

So $\text{FS}[\sin x] = (8/\pi)\sum_{k=1}^{\infty}(-1)^k k/(4k^2 - 9)\sin(2kx/3)$, which differs from that found in part **(a)** because the intervals are different.

5. Just because a series $\sum_n A_n\cos nx + B_n\sin nx$ has the form of a Fourier Series does *not* necessarily mean that it is one. Here we look at three such series which are *not* the Fourier Series of any function in $\text{PC}[-\pi, \pi]$. By the Parseval Relation (Theorem 10.3.6), $\sum\left(1/(\pi n^{1/4})\right)^2$ must converge if $f(x) = \sum n^{-1/4}\cos nx$ is a function in $\text{PC}[-\pi, \pi]$. But $\pi^{-2}\sum n^{-1/2}$ diverges (compare with the divergent harmonic series $\sum n^{-1}$). So, $\sum n^{-1/4}\cos nx$ is not the Fourier Series of any function in $\text{PC}[-\pi, \pi]$.

    Similarly, $g = \sum(\sin n)\sin nx$ is not the Fourier Series of a function in $\text{PC}[-\pi, \pi]$ because $\sum(\sin n)^2$ doesn't converge at all [the n-th term does not $\to 0$ as $n \to \infty$].

    Since $\ln(n) < n^{1/4}$ for n large enough, we have that $1/(\ln(n))^2 > 1/n^{1/2}$ for n large enough [use L'Hôpital's Theorem to show that $\lim_{n\to\infty} \ln(n)/n^{1/4} = 0$]. By the Comparison Test $\sum 1/(\ln(n))^2$ diverges since $\sum 1/n^{1/2}$ does. By Parseval's Theorem (again), $h = \sum(1/\ln(n)) \sin nx$ cannot be the Fourier Series of any function in PC$[-\pi, \pi]$.

7. Let a be any given real number. We want to show that if f has period $2T$, then $\int_{-T}^{T} f(x)\,dx = \int_{a}^{a+2T} f(s)\,ds$. The interval $a \le x < a + 2T$ must contain precisely one point of the lattice $0, \pm T, \pm 3T, \ldots$. Let this point be $(2k+1)T$. Then

$$\int_{a}^{a+2T} f(x)dx = \int_{a}^{(2k+1)T} f(x)dx + \int_{(2k+1)T}^{a+2T} f(x)dx$$

Making the variable substitution $x = t + 2kT$ in the first integral, and the substitution $x = t + (2k+2)T$ in the second integral, we see that

$$\int_{a}^{(2k+1)T} f(x)dx = \int_{a-2kT}^{T} f(t)dt, \quad \int_{(2k+1)T}^{a+2T} f(x)dx = \int_{-T}^{a-2kT} f(t)dt$$

and we have that

$$\int_{a}^{a+2T} f(x)dx = \int_{-T}^{2a-2kT} f(t)dt + \int_{a-2kT}^{T} f(t)dt = \int_{-T}^{T} f(t)dt$$

9. **(a).** By the Totality Theorem (Theorem 10.3.5) if $\{\cos nx : n = 0, 1, 2, \ldots\}$ were a basis for $PC[-\pi, \pi]$, then the *only* element in $PC[-\pi, \pi]$ orthogonal to $\cos nx$, for all $n = 0, 1, 2, \ldots$, is the zero element. But since $\langle \sin x, \cos nx \rangle = 0$, for all $n = 0, 1, 2 \ldots$, it follows that $\{\cos nx : n = 0, 1, 2, \ldots\}$ is *not* a basis for $PC[-\pi, \pi]$.

(c). The set $\{\sin x, \ldots, \sin nx\}$ is a basis for the subspace of all odd functions in PC$[-\pi, \pi]$ since $\langle f, \cos nx \rangle = 0$, for $n = 0, 1, \ldots$ if f is odd. It is *not* a basis of PC$[-\pi, \pi]$ since, for example $\langle 1, \sin nx \rangle = 0$, $n = 1, 2, \ldots$, but 1 is not the zero function.

Background Material: Dirichlet Representation Theorem

THEOREM

> Dirichlet Representation Theorem. Suppose that f is in PC$[-\pi, \pi]$, and that $\tilde{f}$ is the periodic extension of f in $\mathbb{R}$ with period 2π. Suppose that $S_n(x)$ denotes the n-th partial sum of FS$[f]$: $S_n(x) = A_0 + \sum_{k=1}^{n}(A_k \cos kx + B_k \sin kx)$, where A_k, B_k are the Fourier-Euler trigonometric coefficients of f with respect to the basis $\{1, \cos x, \sin x, \ldots, \cos nx, \sin nx, \ldots\}$. Then we have the *Dirichlet Representation* for $S_n(x)$,
>
> $$S_n(x) = \int_{-\pi}^{\pi} \tilde{f}(x+s) \frac{\sin(n+\frac{1}{2})s}{2\pi \sin(s/2)}\,ds, \qquad x \text{ in } \mathbb{R}, \quad n = 0, 1, 2, \ldots \qquad \text{(i)}$$
>
> The function $D_n(s) = \sin[(n+\frac{1}{2})s]/[2\pi \sin(s/2)]$ appearing in the integrand of (i) is called the *Dirichlet kernel* and has the property that
>
> $$\int_{-\pi}^{\pi} D_n(s)\,ds = 1, \qquad \text{for all } n = 0, 1, 2, \ldots \qquad \text{(ii)}$$

To see this, use the definition of the coefficients A_k, B_k, interchanging summation and integration, and the identity $\cos(\alpha - \beta) = \cos \alpha \cos \beta + \sin \alpha \sin \beta$ to obtain

$$S_n(x) = \frac{1}{\pi} \int_{-\pi}^{\pi} f(t) \left[\frac{1}{2} + \sum_{k=1}^{n} \cos k(t - x) \right] dt, \qquad \text{for any } n = 0, 1, 2, \ldots$$

☞ Problem 8 gives a hint on how to show this identity.

Using the identity $\frac{1}{2} + \sum_{k=1}^{n} \cos kx = \sin[(n + \frac{1}{2})s]/[2 \sin(s/2)]$, we see that

$$S_n(x) = \frac{1}{\pi} \int_{-\pi}^{\pi} f(t) \frac{\sin(n + \frac{1}{2})(t - x)}{2 \sin \frac{1}{2}(t - x)} dt \qquad \text{(iii)}$$

Now after replacing f by $\tilde{f}$ in (iii), making the change of variables $t - x = s$, and applying the Integration Theorem for Periodic Functions (Theorem 10.4.2), the desired formula (i) for $S_n(x)$ is obtained. The identity (ii) results immediately from application of (i) to the function $f(x) = 1$, for all x, and we are done.

Background Material: Verification of Theorem 10.4.3

To see this, first write $S_n(x)$ in (i) above as

$$S_n(x) = \int_{-\pi}^{0} \tilde{f}(x + s) D_n(s) \, ds + \int_{0}^{\pi} \tilde{f}(x + s) D_n(s) \, ds$$

We will show that

$$\int_{0}^{\pi} \tilde{f}(x_0 + s) D_n(s) \, ds \to \frac{1}{2} \tilde{f}(x_0^+), \qquad \text{as } n \to \infty$$
$$\int_{-\pi}^{0} \tilde{f}(x_0 + s) D_n(s) \, ds \to \frac{1}{2} \tilde{f}(x_0^-), \qquad \text{as } n \to \infty \qquad \text{(i)}$$

which implies the assertion. We show only the first statement in (i). Note that $D_n(s)$ is an even function, so from (ii), $\int_{0}^{\pi} D_n(s) \, ds = \int_{-\pi}^{0} D_n(s) \, ds = 1/2$. Using this fact and the identity $\sin(\alpha + \beta) = \sin \alpha \cos \beta + \cos \alpha \sin \beta$, we have

$$\int_{0}^{\pi} \tilde{f}(x_0 + s) D_n(s) \, ds - \frac{1}{2} \tilde{f}(x_0^+) = \int_{0}^{\pi} [\tilde{f}(x_0 + s) - \tilde{f}(x_0^+)] D_n(s) \, ds$$

$$= \frac{1}{2\pi} \int_{-\pi}^{\pi} \{H(s)[\tilde{f}(x_0 + s) - \tilde{f}(x_0^+)]\} \cos ns \, ds \qquad \text{(ii)}$$

$$+ \frac{1}{\pi} \int_{-\pi}^{\pi} \left\{ H(s) \cdot \frac{\tilde{f}(x_0 + s) - \tilde{f}(x_0^+)}{s} \frac{s/2}{\sin(s/2)} \cos(s/2) \right\} \sin ns \, ds$$

where $H(s)$ is the piecewise constant function whose value is 1, $0 < s < \pi$, and 0, $-\pi < s < 0$. Now $H(s)[\tilde{f}(x_0 + s) - \tilde{f}(x_0^+)]$ belongs to the class $PC[-\pi, \pi]$ as a function of s, so the Decay of Coefficients Theorem implies that the first integral in (ii) tends to zero as $n \to \infty$. If we can show that the function

$$H(s) \cdot \frac{\tilde{f}(x_0 + s) - \tilde{f}(x_0^+)}{s} \frac{s/2}{\sin(s/2)} \cos(s/2)$$

is also in PC$[-\pi, \pi]$ as a function of s, the Decay of Coefficients Theorem implies that this integral also tends to zero as $n \to \infty$, finishing the proof.

Observe that $H(s)$, $(s/2)\sin(s/2)$, and $\cos(s/2)$ are all piecewise continuous on $-\pi \leq s \leq \pi$. Now since $H(s) = 0$ for $-\pi \leq s < 0$, it remains only to show that $[\tilde{f}(x_0 + s) - \tilde{f}(x_0^+)]/s$ is piecewise continuous on $0 \leq s \leq \pi$. The only trouble point is $s = 0$. Since f is piecewise smooth on $[-\pi, \pi]$, it follows that $\tilde{f}(x_0 + s)$ is piecewise smooth on $-\pi \leq s \leq \pi$. So for all small enough positive s, the Mean Value Theorem implies that

☞ Theorem B.5.4 has a statement of the Mean Value Theorem.

$$\frac{\tilde{f}(x_0 + s) - \tilde{f}(x_0^+)}{s} = \tilde{f}'(x_0 + s^*) \qquad \text{for some } s^* \text{ with } 0 < s^* < s \qquad \text{(iii)}$$

Since $\tilde{f}'(x_0 + s^*) \to \tilde{f}'(x_0^+)$ as $s \to 0^+$, we see from (iii) that the function $[\tilde{f}(x_0 + s) - \tilde{f}(x_0^+)]/s$ has passed the last test to show that it is piecewise continuous on $0 \leq s \leq \pi$, and we are done.

10.5 Half-Range and Exponential Fourier Series

Comments

The Fourier techniques of half-range and complex exponential expansions are "bread-and-butter" methods in doing applications. It is worthwhile pointing out the advantages of using the complex basis functions e^{ikx}: the Fourier-Euler coefficients formulas are a lot easier to remember, and one doesn't have to compute Fourier sine and cosine coefficients separately. The flip side of the coin, however, is that without the often tedious simplification at the end of the process, the Fourier Series of a real function may appear to be complex-valued.

1. We are looking for Fourier Sine Series here.

    (a). The function $f(x) = 1$.

    $$\text{FSS}[1] = \sum_{k=0}^{\infty} b \sin kx, \qquad b_k = \frac{2}{\pi} \int_0^{\pi} 1 \cdot \sin kx \, dx$$

    So $b_k = -2(\cos kx)/k\pi \Big|_{x=0}^{x=\pi} = 2[1 - \cos k\pi]/k\pi$, and we have $b_k = 0$ if k is even, and $b_k = 4/k\pi$ if k is odd. Then

    $$\text{FSS}[1] = \frac{4}{\pi} \sum_{\text{odd } k} \frac{1}{k} \sin kx$$

    We have FCS[1]=FS[1]=1 since $\Phi_c = \{1, \cos x, \ldots, \cos nx, \ldots\}$.

3. We suppose that f is in PC[0, c] and that $f(x) = f(c - x)$ for $c/2 < x < c$.

    (a). Let's look at FSS[f]. For $k = 2m$, $m = 1, 2, \ldots$, we see that

    $$\int_0^c f(x) \sin \frac{2m\pi x}{c} \, dx = \int_0^{c/2} f(x) \sin \frac{2m\pi x}{c} \, dx + \int_{c/2}^c f(x) \sin \frac{2m\pi x}{c} \, dx \qquad \text{(i)}$$

Making the substitution $x = c - y$ in the first integral on the right side of (i) and using an identity from Appendix B.4 for $\sin(\alpha - \beta)$, we see that

$$\int_0^{c/2} f(x) \sin \frac{2m\pi x}{c} \, dx = -\int_c^{c/2} f(c - y) \sin \frac{2m\pi(c - y)}{c} \, dy$$

$$= -\int_{c/2}^{c} f(y) \sin \frac{2m\pi y}{c} \, dy$$

The right side of (i) adds up to zero, and we have

$$\int_0^c f(x) \sin \frac{2m\pi x}{c} \, dx = 0, \qquad m = 1, 2, \ldots$$

This leaves only the terms $\sin[(2k + 1)\pi x/c]$ in FSS[f].

5. The circuit equation is $0.5q'' + 10q' + 10^4 q = E(t)$. From the hint we see that FS[E] $= \pi/2 + \pi^{-1} \sum_{k=1}^{\infty} (e^{ikt} + e^{-ikt})/(4k^2)$. Now if we put FS[$q$] $= \sum_{k=-\infty}^{\infty} c_k e^{ikt}$, then FS[$q'$] $= \sum_{k=-\infty}^{\infty} ik c_k e^{ikt}$, and FS[$q''$] $= \sum_{k=-\infty}^{\infty} (ik)^2 c_k e^{ikt}$. We can combine these results to compute FS[$0.5q'' + 10q' + 10^4 q$] and equate it to FS[E], and by comparing coefficients we obtain that

$$[0.5(ik)^2 + 10ik + 10^4]c_k = \begin{cases} \pi/2, & k = 0, \quad k \text{ odd} \\ -1/(2m^2), & k \neq 0, \quad k = 2m \end{cases}$$

Since the coefficient of c_k in the equation above is not zero for any k, we can solve the equation for c_k uniquely and if these values are inserted into $\sum_{k=-\infty}^{\infty} c_k e^{ikx}$, this series converges to the steady-state solution of the given circuit equation.

10.6 Sturm-Liouville Problems

Comments

The separation step in the technique of separation of variables for a boundary-initial value problem for a PDE always generates a Sturm-Liouville problem. This typically involves solving an ordinary differential operator equation $Ly = \lambda y$, subject to some conditions on the solution $y(x)$ at $x = a$ and $x = b$, say. More precisely, the problem is to find the values of λ so that the problem $Ly = \lambda y$ with boundary conditions has a nontrivial solution. We must find the eigenvalues of the operator L whose domain is restricted by the boundary conditions. The Sturm-Liouville Theory provides the mathematical foundations for all of this. Everything needed in Sections 10.7–10.9 and the ideas that underlie what is in Section 11.8 for the separation step in solving PDE problems can be found right here.

1. (a). $L[y] = y''$, Dom(L) $= \{y$ in $C^2[0, \pi/2]: y(0) = y(\pi/2) = 0\}$. From the Nonpositivity of Eigenvalues Theorem (Theorem 10.6.3) all the eigenvalues of L are nonpositive. First try $\lambda = 0$. The eigenfunctions (if any) must have the form $y = Ax + B$. The endpoint conditions imply that $A = B = 0$, so $\lambda = 0$ cannot be an eigenvalue. Now try $\lambda = -k^2$,

$k \geq 0$. The eigenfunctions (if any) must have the form $y = A \sin kx + B \cos kx$. The condition $y(0) = 0$ implies that $B = 0$. The condition $y(\pi/2) = 0$ implies that $\sin(k\pi/2) = 0$. This can only happen if $k\pi/2 = n\pi$, $n = 1, 2, \ldots$. So $\lambda_n = -4n^2$ is an eigenvalue for each $n = 1, 2, \ldots$, and the corresponding eigenspace V_{λ_n} has $y_n = \sin 2nx$ as a basis. The Orthogonality of Eigenspaces Theorem 10.6.2 implies that the eigenspaces V_{λ_1}, $V_{\lambda_2}, \ldots$ are orthogonal with respect to the standard scalar product in $PC[0, \pi/2]$ and that $\Phi = \{\sin 2x, \cdots, \sin 2nx, \cdots\}$ is a basis for $PC[0, \pi/2]$.

(c). $L[y] = y''$, $\text{Dom}(L) = \{y \text{ in } \mathbf{C}^2[0, T]: y'(0) = y'(T) = 0\}$. First, $\lambda = 0$ is an eigenvalue and $y = 1$ a corresponding eigenvector. As in part **(a)**, the other eigenvalues must be negative, $\lambda = -k^2$, and $y_k = A_k \cos kx + B_k \sin kx$. Imposing the boundary conditions, we see that $B_k = 0$, and the eigenvalues are $\lambda_n = -(n\pi/T)^2$, $n = 0, 1, 2, \ldots$ with each corresponding eigenspace V_{λ_n} spanned by the eigenfunction $y_n = \cos(n\pi x/T)$. The eigenspaces V_{λ_n} are orthogonal with respect to the standard scalar product in $PC[0, T]$, and the orthogonal set $\Phi = \{\cos(n\pi x/T): n = 0, 1, 2, \ldots\}$ is a basis for $PC[0, T]$.

(e). $L[y] = y''$, $\text{Dom}(L) = \{y \text{ in } \mathbf{C}^2[0, \pi]: y(0) = y(\pi) + y'(\pi) = 0\}$. First, find eigenvalues of L of the form $\lambda = k^2$, for some $k > 0$. Such eigenvalues would have eigenfunctions (if any) of the form $y = A \sinh kx + B \cosh kx$. The condition $y(0) = 0$ implies that $B = 0$. The condition $y(\pi) + y'(\pi) = 0$ implies that $\sinh k\pi + k \cosh k\pi = 0$ or $\tanh r = -\pi/r$, where $k\pi = r > 0$. The sketch in Fig. 1(e), Graph 1 indicates that there is no value $r > 0$ such that $\tanh r = -\pi/r$, so no positive eigenvalues. (Note that the Nonpositivity of Eigenvalues Theorem 10.6.3 would have given the result immediately). Next try $\lambda = 0$. The eigenfunctions (if any) must have the form $y = Ax + B$. The endpoint conditions imply that $A = B = 0$, so $\lambda = 0$ cannot be an eigenvalue. Finally, let's try $\lambda = -k^2$, $k \geq 0$. In this case any eigenfunction would have the form $y = A \cos kx + B \sin kx$. The condition $y(0) = 0$ implies that $A = 0$, and the other boundary condition implies that $\sin k\pi + k \cos k\pi = 0$ or $\tan r = -\pi/r$, for $r > 0$, where $k\pi = r$. The sketch in Fig. 1(e), Graph 2 indicates that there is an infinite sequence $r_1, r_2, \ldots$, for which $\tan r_n = -\pi r_n$, $r_n > 0$. Note that $(n - 1/2)\pi < r_n < n\pi$, for all n, and in fact $r_n \to (n + 1/2)\pi$ as $n \to \infty$. So $\lambda_n = -k_n^2 = -(r_n/\pi)^2$ are the eigenvalues of L, and the corresponding eigenspace V_{λ_n} is spanned by $y_n = \sin k_n x = \sin(r_n x/\pi)$. Because L is a symmetric operator, the eigenspaces are orthogonal with the standard scalar product in $PC[0, \pi]$. Finally, the Sturm-Liouville Theorem 10.6.4 implies that $\Phi = \{\sin(r_n x/\pi): n = 1, 2, \ldots\}$ is a basis for $PC[0, \pi]$.

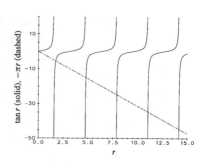

Problem 1(e), Graph 1.   Problem 1(e), Graph 2.

10.7 Separation of Variables

Comments

Now we can approach a boundary/initial value problem for a PDE in a systematic way, using the techniques introduced earlier. A systematic approach is very helpful because there are so many separate steps in constructing the series solution of the problem that a "do-it-by-the-numbers" approach is preferred. This does not mean doing a problem without thought, because every new problem brings new geometry, data sets, or coefficients in the PDE. However, the template we have arranged in the form of seven steps (numbered as I–VII in the text) helps in avoiding needless errors and pointless searching for what to do next. The whole process has at its core the Method of Eigenfunction Expansions, and we make note of that important fact. Finally, we show how to solve boundary/initial value problems for PDEs when the PDE, the boundary conditions, or both, are not homogeneous. Here is where eigenfunction expansions play a central role in constructing the solution. We distinguish between a "formal solution" of a boundary/initial value problem where we find the solution as a series but don't worry about the convergence, from a "classical solution" where we add enough extra smoothness conditions on the data that all the series converge to the desired sum. On a first "read-through" you might ignore the distinction and focus on the formal solution. You might also skip the verification of the theorems.

1. The problems all refer to the guitar string problem (1). The initial shape $f(x)$ and velocity $g(x)$ are given.

   (a). The formal solution of this problem takes the form (8) with coefficients given by (9). Using the given data $f(x) = 0$, $g(x) = 3\sin(\pi x/L)$, we see that $A_n = 0$ for all n, and that $B_1 = (2/\pi c)\int_0^L 3(\sin(\pi x/L))^2 \, dx = 3L/\pi c$ and $B_n = 0$ for $n \geq 2$. The solution of this problem is

   $$u(x, t) = \frac{3L}{\pi c}\sin\frac{\pi ct}{L}\sin\frac{\pi x}{L}$$

   (c). Here $f(x) = x(L - x) = -g(x)$. Using (8) and (9), we have that for $n \geq 1$,

   $$A_n = \frac{2}{L}\int_0^L x(L - x)\sin\frac{n\pi x}{L}dx = \frac{4L^2}{n^3\pi^3}(1 - (-1)^{n+1})$$

and

$$B_n = \frac{2}{n\pi c} \int_0^L x(x-L) \sin \frac{n\pi x}{L} dx = \frac{4L^3}{n^4\pi^4 c}((-1)^{n+1} - 1) = \begin{cases} -8L^3/(n^4\pi^4 c), & \text{n even} \\ 0, & \text{n odd} \end{cases}$$

Substituting these values for A_n, B_n in the series for $u(x, t)$ in (8) yields the solution.

3. A separated solution of the PDE $u_{tt} + b^2 u_t - a^2 u_{xx} = 0$ has the form $u = X(x)T(t)$ where

$$X(x)T''(t) + b^2 X(x)T'(t) - a^2 X''(x)T(t) = 0$$

Dividing by $a^2 XT$ and separating, we have that

$$\frac{X''}{X} = \frac{T'' + b^2 T'}{a^2 T} = \lambda$$

where λ is a separation constant. Imposing the boundary conditions we have the Sturm-Liouville Problem

$$X'' = \lambda X, \quad X(0) = X(L) = 0$$

and the ODE

$$T'' + b^2 T' - \lambda a^2 T = 0$$

We have that $X_n = C_n \sin(n\pi x/L)$, $\lambda_n = -n^2\pi^2/L^2$, $n = 1, 2, \ldots$, exactly as in (7). Recalling that we require that $b^2 < 2\pi a/L$, then for all integers $n = 1, 2, \ldots$, the roots of the quadratic $r^2 + b^2 r - \lambda_n a^2$ are the complex conjugates $-b^2/2 \pm i\beta_n$, where $\beta_n = \sqrt{-\lambda_n a^2 - b^4/4}$. Then we have that

$$T_n(t) = e^{-b^2 t/2} [A_n \cos \beta_n t + B_n \sin \beta_n t]$$

and the solution has the form

$$u(x, t) = e^{-b^2 t/2} \sum_1^\infty \sin \frac{n\pi}{L} x [A_n \cos \beta_n t + B_n \sin \beta_n t]$$

Now set $t = 0$ to obtain

$$u(x, 0) = f(x) = \sum_1^\infty A_n \sin \frac{n\pi}{L} x$$

and

$$u_t(x, 0) = 0 = \sum_1^\infty \sin \frac{n\pi}{L} x \left[-\frac{b^2 A_n}{2} + \beta_n B_n \right]$$

We have that

$$A_n = \frac{2}{L} \int_0^L f(x) \sin(n\pi x/L) \, dx$$

$$B_n = b^2 A_n/(2\beta_n)$$

and we are done.

5. The problem is $u_{tt} - u_{xx} = 0$, $0 < x < 1$, $t > 0$; $u(0, t) = \sin(\pi t/2)$, $u(1, t) = 0$, $t \geq 0$; $u(x, 0) = f(x)$, $u_t(x, 0) = g(x)$, $0 \leq x \leq 1$. We need to shift the boundary data $\sin(\pi t/2)$ onto an initial condition. Recall that $X(x)T(t)$ is a solution of the wave equation $u_{tt} = u_{xx}$ if $X'' = \lambda X$ and $T'' = \lambda T$ for the same constant λ.

(a). Let $v(x, t)$ satisfy the given conditions of the "intermediate" problem, $v_{tt} - v_{xx} = 0$, $v(0, t) = \sin \pi t/2$, $v(1, t) = 0$, $t \geq 0$. Let $u(x, t)$ satisfy the PDE and the conditions of the original problem. Then $w(x, t) = u - v$ satisfies the wave equation $w_{tt} - w_{xx} = 0$, and the boundary conditions

$$w(0, t) = u(0, t) - v(0, t) = \sin \frac{\pi t}{2} - \sin \frac{\pi t}{2} = 0$$

$$w(1, t) = u(1, t) - v(1, t) = 0 - 0 = 0$$

for $t \geq 0$. The initial conditions satisfied by w are $w(x, 0) = u(x, 0) - v(x, 0) = f(x) - v(x, 0)$, and $w_t(x, 0) = u_t(x, 0) - v_t(x, 0) = g(x) - v_t(x, 0)$.

(c). Note that $w(x, t) = u(x, t) - \cos(\pi x/2) \sin(\pi t/2)$ satisfies the conditions

$$
\begin{aligned}
w_{tt} - w_{xx} &= 0, & 0 < x < 1, \quad t > 0 \\
w(0, t) = w(1, t) &= 0, & t > 0 \\
w(x, 0) &= f(x), & 0 < x < 1 \\
w_t(x, 0) &= -(\pi/2) \cos(\pi x/2), & 0 < x < 1
\end{aligned}
$$

Following (9), we see that w has the form

$$w(x, t) = \sum_{n=1}^{\infty} \sin n\pi x (A_n \cos n\pi t + B_n \sin n\pi t)$$

where for $n \geq 1$

$$A_n = 2 \int_0^1 f(x) \sin n\pi x \, ds$$

$$B_n = \frac{2}{n\pi} \int_0^1 \left(-\frac{\pi}{2} \cos \frac{\pi x}{2} \right) \sin n\pi x \, dx$$

$$= -\frac{1}{2n} \int_0^1 (\sin(n + 1/2)\pi x + \sin(n - 1/2)\pi x) \, dx = -\frac{4}{\pi(4n^2 - 1)}$$

The solution u of the boundary/initial value problem [with $u_t(x, 0) = 0$] is

$$u(x, t) = \sum_{n=1}^{\infty} \sin n\pi x (A_n \cos n\pi t + B_n \sin n\pi t) + \cos(\pi x/2) \sin(\pi t/2)$$

where A_n and B_n are as given above.

7. We want to find a periodic function $\mu(t)$ to break the string modeled by $u_{tt} - c^2 u_{xx} = 0$, $0 < x < L$, $t > 0$; $u(0, t) = 0$, $u(L, t) = \mu(t)$, $t \geq 0$; $u(x, 0) = 0$, $u_t(x, 0) = 0$, $0 \leq x \leq L$. Consider the "shaking" function $\mu(t) = \cos \omega t$ for some constant ω. To solve the boundary/initial value problem for u, we first solve the "intermediate" problem

$$v_{tt} - c^2 v_{xx} = 0, \qquad 0 < x < L, \quad t > 0$$
$$v(0, t) = 0, \qquad t > 0$$
$$v(L, t) = \cos \omega t, \qquad t > 0$$

Now the separated function $X(x)T(t)$ solves the wave equation if $X'' = \lambda X$ and $T'' = c^2 \lambda T$ for some constant λ. Say we search for v in the form $v(x, t) = X(x) \cos \omega t$. Then clearly $X(x)$ must be a solution of $X'' = -(\omega/c)^2 X$ and satisfy the conditions $X(0) = 0$, $X(L) = 1$. Observe that if $\omega L/c \neq n\pi$, then

$$X(x) = \left(\sin \frac{\omega L}{c} \right)^{-1} \sin \frac{\omega x}{c}$$

has this property. The intermediate problem has the solution

$$v(x, t) = \left(\sin \frac{\omega L}{c} \right)^{-1} \sin\left(\frac{\omega x}{c}\right) \cos \omega t$$

Now we put $w = u - v$, where u solves the original problem and v the intermediate problem, and observe that w solves the problem

$$w_{tt} - w_{xx} = 0, \qquad\qquad 0 < x < L, \quad t \geq 0$$
$$w(0, t) = w(L, t) = 0, \qquad t > 0$$
$$w(x, 0) = -(\sin(\omega L/c))^{-1} \sin(\omega x/c), \qquad 0 < x < L$$
$$w_t(x, 0) = 0, \qquad\qquad 0 < x < L$$

The solution of this problem for w is given by (9). Note that $B_n = 0$ for all n since $g(x) = 0$. To find the coefficients A_n, we must evaluate the integrals

$$\frac{2}{L} \int_0^L \sin \frac{\omega x}{c} \sin \frac{n\pi x}{L} dx = \frac{1}{L} \int_0^L \left\{ \cos\left(\frac{\omega}{c} - \frac{n\pi}{L}\right) x - \cos\left(\frac{\omega}{c} + \frac{n\pi}{L}\right) x \right\} dx$$

$$= \frac{(-1)^n 2 n\pi c^2}{(\omega L)^2 - (n\pi c)^2} \sin \frac{\omega L}{c} = \alpha_n$$

Then the coefficient A_n is

$$A_n = -\left(\sin \frac{\omega L}{c} \right)^{-1} \alpha_n = \frac{(-1)^{n+1} 2 n\pi c^2}{(\omega L)^2 - (n\pi c)^2}$$

The solution to the problem for w is

$$w(x, t) = \sum_{n=1}^{\infty} \frac{(-1)^{n+1} 2 n\pi c^2}{(\omega L)^2 - (n\pi c)^2} \sin \frac{n\pi x}{L} \cos \frac{n\pi ct}{L}$$

so $u = w + v$ is the solution of the original problem. If we allow ω to approach the fundamental frequency $\pi c/L$, then

$$\left(\frac{\omega L}{c} \right)^2 - \pi^2 \to 0$$

and both $v(x, t)$ and the first term of $w(x, t)$ become unbounded (i.e., the string "breaks").

10.8 The Heat Equation: Optimal Depth for a Wine Cellar

Comments

Wave motion has been the physical phenomenon supporting our discussion of PEs until now. The conduction of heat is shown in this section to be modeled by a linear PDE as well, but it is not the wave equation. The heat (or diffusion) PDE is derived from physical principles concerning the flow of thermal energy from warmer to cooler regions in a body. Natural boundary conditions are introduced that model heat sources or sinks or insulating material placed around the boundary of the body. Separation of variables is applied to construct a series solution for the flow of heat (measured by changing temperatures) in a long circular rod whose lateral walls are wrapped in thermal insulation. Then we show how to determine the optimal depth of a storage cellar to protect its contents from the temperature variations at the earth's surface. Finally, we compare and contrast properties of solutions of the heat equation with those of solutions of the wave equation. In particular, the heat operator smooths out any "corners" or discontinuities in the initial heat curve, while the wave equation simply propagates the initial kinks and jumps. Moreover, thermal energy moves "infinitely" fast, while a signal propagates with a finite speed. Some of the properties, incidentally, show that the wave and heat equations cannot be exact models of the corresponding real phenomena. We like to present the material of Sections 10.7 and 10.8 back-to-back because the models are so "physical" and lead to such contrasting conclusions.

---

1. We have shown that $X(x)T(t)$ is a solution of the heat equation: $u_t - Ku_{xx} = 0$ if $X'' = \lambda X$ and $T' = \lambda KT$ for the same constant λ. In parts **(a)** and **(c)** of this problem we require that $u(0, t) = u(L, t) = 0$. To save space, the details are only carried out for **(a)**, **(e)**, and **(g)**. [*Hint*: See the subsection of 10.8, Temperature in a Rod.]

    (a). Let us consider the separated solution $= X(x)T(t)$. Inserting $X(x)T(t)$ into the PDE and applying the boundary conditions, we see that $X(x)$ and $T(t)$ must be solutions of

    $$\frac{d^2 X}{d^2 x} - \lambda X = 0, \quad X(0) = 0, \quad X(L) = 0$$
    $$\frac{dT}{dt} - \lambda KT = 0$$

    As in (20) and (21) in the text, we have that $\lambda_n = -(n\pi/L)^2$, $X_n(x) = A_n \sin(n\pi x/L)$, $T_n(t) = B_n \exp[-K(n\pi/L)^2 t]$, $n = 1, 2, \ldots$, where A_n and B_n are any constants. The solutions of the PDE, $u_t - Ku_{xx} = 0$ and the boundary conditions $u(0, t) = u(L, t) = 0$ are given by

    $$u_n = X_n(x)T_n(t) = C_n \sin\frac{n\pi x}{L} \exp\left[-K\left(\frac{n\pi}{L}\right)^2 t\right], \qquad n = 1, 2, \ldots$$

    where $C_n = A_n B_n$ is an arbitrary constant. Since the PDE and the boundary conditions are homogeneous in this problem,

$$u(x, t) = \sum_{n=1}^{\infty} C_n \sin \frac{n\pi x}{L} \exp\left[-K\left(\frac{n\pi}{L}\right)^2 t\right]$$

is also a solution of the PDE and the boundary conditions. Now we only need to determine the constants C_n so that $u(x, t)$ satisfies the initial condition. That is, we must choose C_n so that

$$u(x, 0) = \sum_{n=1}^{\infty} C_n \sin \frac{n\pi x}{L} = \sin \frac{2\pi x}{L}, \quad 0 \le x \le L$$

So $C_1 = C_3 = \cdots = 0$ and $C_2 = 1$. The solution is

$$u(x, t) = \sin \frac{2\pi x}{L} \exp\left[-K\left(\frac{2\pi}{L}\right)^2 t\right]$$

This represents the temperature as one hump of a sine curve whose amplitude decays exponentially in time to zero.

(c). Since $f(x) = u_0$,

$$C_n = \frac{2}{L} \int_0^L u_0 \sin \frac{n\pi x}{L} dx = \frac{2u_0}{n\pi}[1 - (-1)^n]$$

The series solution is

$$u(x, t) = \frac{2u_0}{\pi} \sum_{n=1}^{\infty} \frac{1 - (-1)^n}{n} \sin \frac{n\pi x}{L} \exp\left[-K\left(\frac{n\pi}{L}\right)^2 t\right]$$

$$= \frac{4u_0}{\pi} \sum_{\text{odd } n} \frac{1}{n} \sin \frac{n\pi x}{L} \exp\left[-K\left(\frac{n\pi}{L}\right)^2 t\right]$$

Note that for all $t \ge 0$ the series solution for u is zero at $x = 0, L$. As time goes on, the temperatures everywhere decline to the zero values maintained at the endpoints.

(e). Here the boundary conditions are $u(0, t) = 0$ and $u_x(L, t) = 0$ [perfect insulation at $x = L$]. Let's consider the separated solution $= X(x)T(t)$. Inserting $X(x)T(t)$ into the PDE and boundary conditions, we see that $X(x)$ and $T(t)$ must be solutions of

$$X''(x) - \lambda X(x) = 0, \quad X(0) = 0, \quad X'(L) = 0$$
$$T'(t) - \lambda K T(t) = 0$$

We have $\lambda_n = -[(2n - 1)\pi/(2L)]^2$ and

$$X_n(x) = A_n \sin \frac{(2n - 1)\pi x}{2L}, \quad T_n(t) = B_n \exp\left[-K\left(\frac{2n - 1}{2L}\pi\right)^2 t\right], \quad n = 1, 2, \ldots$$

where A_n and B_n are any constants. The separated solutions of the PDE, $u_t - K u_{xx} = 0$, and the boundary conditions $u(0, t) = u_x(L, t) = 0$ are given by

$$u_n = X_n(x)T_n(t) = C_n \sin \frac{(2n-1)\pi x}{2L} \exp\left[-K\left(\frac{2n-1}{2L}\pi\right)^2 t\right], \qquad n = 1, 2, \ldots$$

where $C_n = A_n B_n$ is an arbitrary constant. Since the PDE and the boundary conditions are homogeneous in this problem,

$$u(x, t) = \sum_{n=1}^{\infty} C_n \sin \frac{(2n-1)\pi x}{2L} \exp\left[-K\left(\frac{2n-1}{2L}\pi\right)^2 t\right]$$

is also a solution of the PDE and the boundary conditions. Now we only need to determine constants C_n so that $u(x, t)$ satisfies the initial condition. That is, we must choose C_n so that

$$u(x, 0) = \sum_{n=1}^{\infty} C_n \sin \frac{(2n-1)\pi x}{2L} = \sin \frac{\pi x}{2L}, \qquad 0 \le x \le L$$

We have $C_1 = 1$, $C_2 = C_3 = \cdots = 0$. The solution is

$$u(x, t) = \sin \frac{\pi x}{2L} \exp\left[-K\left(\frac{\pi}{2L}\right)^2 t\right]$$

and we have the initial temperature profile decaying exponentially to 0. Note that the condition $u_x(L, t) = 0$ means that a perfect insulator is clamped to the far end of the rod.

(g). Here we have perfect insulation at both ends: $u_x(0, t) = u_x(L, t) = 0$. Let us consider the separated solution $= X(x)T(t)$. Inserting $X(x)T(t)$ into the PDE and the boundary conditions, we see that $X(x)$ and $T(t)$ must be solutions of

$$\frac{d^2 X}{d^2 x} - \lambda X = 0, \quad X'(0) = 0, \quad X'(L) = 0$$

$$\frac{dT}{dt} - \lambda K T = 0$$

We have $\lambda_n = -(n\pi/L)^2$ and

$$X_n(x) = A_n \cos \frac{n\pi x}{L}, \qquad T_n(t) = B_n \exp[-K(\frac{n\pi}{L})^2 t], \qquad n = 0, 1, 2, \ldots$$

where A_n and B_n are any constants. The separated solutions of the PDE: $u_t - K u_{xx} = 0$ with boundary conditions $u_x(0, t) = u_x(L, t) = 0$ are given by

$$u_n = X_n(x)T_n(t) = C_n \cos\left(\frac{n\pi x}{L}\right) \exp\left[-K\left(\frac{n\pi}{L}\right)^2 t\right], \qquad n = 0, 1, 2, \ldots$$

where $C_n = A_n B_n$ is an arbitrary constant. Since the PDE and the boundary conditions are homogeneous in this problem,

$$u(x, t) = C_0 + \sum_{n=1}^{\infty} C_n \cos\left(\frac{n\pi x}{L}\right) \exp\left[-K\left(\frac{n\pi}{L}\right)^2 t\right]$$

is also a solution of PDE. Now we only need to determine the constants C_n so that $u(x, t)$ satisfies the initial condition. That is, we must choose C_n so that

$$u(x, 0) = \sum_{n=0}^{\infty} C_n \cos \frac{n\pi x}{L} = x, \quad 0 \le x \le L$$

Therefore, the coefficients C_n are given by

$$C_n = \frac{\langle f, \cos \frac{n\pi x}{L} \rangle}{\| \cos \frac{n\pi x}{L} \|^2} = \begin{cases} \dfrac{1}{L} \displaystyle\int_0^L x\,dx = \dfrac{L}{2}, & n = 0 \\[2mm] \dfrac{2}{L} \displaystyle\int_0^L x \cos \frac{n\pi x}{L} dx = \dfrac{2L}{n^2\pi^2}[(-1)^n - 1], & n > 0 \end{cases}$$

The series solution is

$$u(x, t) = \frac{L}{2} - \frac{4L}{\pi^2} \sum_{\text{odd } n} \frac{1}{n^2} \cos\left(\frac{n\pi x}{L}\right) \exp\left[-K\left(\frac{n\pi}{L}\right)^2 t\right]$$

Note that since both ends of the rod and the cylindrical walls are perfectly insulated, no thermal energy can escape. This means that as time goes on the energy tends towards a level equilibrium. In fact, as $t \to +\infty$ the temperature tends to the uniform value of $L/2$.

3. **(a).** It is easy to show by a direct calculation (given V and U as defined below) that $u(x, t) = V(x, t) + U(x, t)$ satisfies the PDE, $u_t - Ku_{xx} = 0$, $0 < x < 1$, $t > 0$, the boundary conditions $u(0, t) = g_1(t)$ and $u(1, t) = g_2(t)$, $t \ge 0$, and the initial condition $u(x, 0) = f(x)$, $0 \le x \le 1$. Here $V = g_1(t) + [g_2(t) - g_1(t)]x$ and $U(x, t)$ satisfies the equations $U_t - KU_{xx} = -KV_t$, $0 < x < 1$, $t > 0$; $U(0, t) = U(1, t) = 0$, $t \ge 0$; $U(x, 0) = f(x) - V(x, 0)$, $0 \le x \le 1$.

(c). By part **(a)** $u(x, t) = V(x, t) + U(x, t)$ where $V(x, t) = g(t) + [g_2(t) - g_1(t)]x$ and $U(x, t)$ satisfies a certain boundary/initial value problem. In this case $g_1 = \sin t$ and $g_2 = 0$, so $V = (1 - x) \sin t$. The formula for $V(x, t)$ is derived in part **(b)** [not shown here]. The final formula turns out to be

$$u(x, t) = (1 - x)\sin t + \sum_{n=1}^{\infty} e^{-K(n\pi)^2 t}\left[2\int_0^1 f(x)\sin n\pi x\,dx - \frac{2K}{n\pi}\int_0^t \cos\tau\, e^{K(n\pi)^2\tau}d\tau\right]\sin n\pi x$$

5. This problem relates to the wine cellar model described in the text.

(a). Let $u = U + T_0$, where $u(x, t)$ satisfies system (25) in the text. Then, we have

$$\begin{aligned} U_t - KU_{xx} &= 0, & 0 < x < \infty, & \quad -\infty < t < \infty \\ U(0, t) &= A_0 \cos \omega t, & -\infty < t < \infty, & \\ |U(x, t)| &< C, & 0 \le x < \infty, & \quad -\infty < t < \infty \end{aligned}$$

since the surface temperature is given as $u(0, t) = T_0 + A_0 \cos \omega t$. As in the text, we have

$$U = \text{Re}\{A_0 e^{-\alpha x} e^{i(\omega t - \alpha x)}\} = A_0 e^{-\alpha x} \cos(\omega t - \alpha x), \quad \alpha = \left(\frac{\omega}{2K}\right)^{1/2}$$

The temperature at depth x at time t is

$$u = T_0 + A_0 e^{-\alpha x} \cos(\omega t - \alpha x), \quad \omega = \frac{2\pi}{3.15 \times 10^7}, \quad \alpha = \left(\frac{\omega}{2K}\right)^{1/2}$$

The optimal depth x_0 reverses the "seasons" in the wine cellar in comparison to the natural seasons at the surface (see the text). We should have $\alpha x = \pi$:

$$x_0 = \frac{\pi}{\alpha} = \pi \left(\frac{2K}{\omega}\right)^{1/2} \approx 4.45 \text{ m}$$

If, however, ω corresponds to 1 day = 86400 seconds instead of 1 year, then, $\omega = 2\pi/86400$. The optimal depth is

$$x_0 = \frac{\pi}{\alpha} = \pi (0.004)^{1/2} \left(\frac{86400}{2\pi}\right)^{1/2} \approx 23.3 \text{ cm}$$

(c). If the surface temperature wave has the form $T_0 + A_1 \cos \omega_1 t + A_2 \cos \omega_2 t$, let $u = T_0 + U + V$, where

$$
\begin{array}{llll}
U_t - KU_{xx} = 0, & V_t - KV_{xx} = 0, & 0 < x < \infty, & -\infty < t < \infty \\
U(0, t) = A_1 \cos \omega_1 t, & V(0, t) = A_2 \cos \omega_2 t, & -\infty < t < \infty, & \\
|U(x, t)| < C, & |V(x, t)| < C, & 0 \le x < \infty, & -\infty < t < \infty
\end{array}
$$

Similar to part **(a)**, we have

$$U = \text{Re}\{A_1 e^{-\alpha_1 x} e^{i(\omega_1 t - \alpha_1 x)}\} = A_1 e^{-\alpha_1 x} \cos(\omega_1 t - \alpha_1 x), \qquad \alpha_1 = \left(\frac{\omega_1}{2K}\right)^{1/2}$$

$$V = \text{Re}\{A_2 e^{-\alpha_2 x} e^{i(\omega_2 t - \alpha_2 x)}\} = A_2 e^{-\alpha_2 x} \cos(\omega_2 t - \alpha_2 x), \qquad \alpha_2 = \left(\frac{\omega_2}{2K}\right)^{1/2}$$

The temperature at depth x at time t is

$$u = T_0 + A_1 e^{-\alpha_1 x} \cos(\omega_1 t - \alpha_1 x) + A_2 e^{-\alpha_2 x} \cos(\omega_2 t - \alpha_2 x)$$

$$\omega_1 = \frac{\omega_2}{365} = \frac{2\pi}{3.15 \times 10^7}, \qquad \alpha_1 = \frac{\alpha_2}{\sqrt{365}} = \left(\frac{\omega_1}{2K}\right)^{1/2}$$

7. The problem is $u_t - u_{xx} = 0$, $0 < x < 1$, $t > 0$; $u(0, t) = te^{-t}, t \ge 0$; $u(1, t) = 0$, $t \ge 0$; $u(x, 0) = 0.01x(1 - x)$, $0 \le x \le 1$. By the Maximum Principle for the Heat Equation (Theorem 10.8.1), we know that $|u(x, t)| \le \max |u(\bar{x}, \bar{t})|$, where $(\bar{x}, \bar{t})$ is a point on the boundary of the region $R_t = \{(x, \bar{t}) : 0 < x < 1, \ 0 \le \bar{t} \le t\}$. Because $\max |u(\bar{x}, \bar{t})| = \max\{0, 0.0025, e^{-1}\} = 1/e$, we have that $|u(x, t)| \le 1/e$.

10.9 Laplace's Equation

Comments

Laplace's PDE and its solutions, called harmonic functions, have properties quite different from anything we have seen so far with the wave and heat equations and their solutions. We illustrate these points by solving a boundary value problem for steady temperatures in a disk, using circular harmonics in the process. Then, we discuss the averaging and maximal properties of harmonic functions and show that solutions of boundary value problems for Laplace's equation are unique (if there is a solution) and continuous in the data.

1. **(a).** The problem is $u_{xx} + u_{yy} = 0$, $0 < x < L$, $0 < y < M$; $u(0, y) = \alpha(y)$, $u(L, y) = 0$, $0 \leq y \leq M$; $u(x, 0) = u(x, M) = 0$, $0 \leq x \leq L$. The technique of separation of variables can be followed just as in previous sections to construct separated solutions $X_n(x)Y_n(y)$, where

$$\frac{-X_n''}{X_n} = \frac{Y_n''}{Y_n} = \lambda_n, \qquad Y_n(0) = Y_n(M) = 0$$

This implies that

$$\lambda_n = -(n\pi/M)^2, \quad Y_n = A_n \sin(n\pi y/m), \quad X_n = B_n \sinh(n\pi x/M) + C_n \cosh(n\pi x/M)$$

We also use the formula $\sinh(\alpha + \beta) = \sinh\alpha\cosh\beta + \cosh\alpha\sinh\beta$. Finally, we obtain

$$u_\alpha(x, y) = \sum_{n=1}^{\infty} A_n \sin\frac{n\pi y}{M} \sinh\frac{n\pi}{M}(L - x)$$

which satisfies the PDE and the boundary conditions. Since $u_\alpha(0, y) = \alpha(y)$, we need to choose the coefficients A_n so that the Fourier-Euler coefficient formula holds,

$$A_n \sinh\frac{n\pi L}{M} = \frac{2}{M}\int_0^M \alpha(y)\sin\frac{n\pi y}{M}dy$$

(c). Replace the condition $u(0, y) = \alpha(y)$ in part **(a)** by an insulation condition $u_x(0, y) = \tilde{\alpha}(y)$, $0 \leq y \leq M$. We know that the solution of $u_{xx} + u_{yy} = 0$ with boundary conditions

$$u_x(0, y) = \tilde{\alpha}(y), \qquad u(L, y) = 0, \quad 0 \leq y \leq M$$
$$u(x, 0) = 0, \qquad u(x, M) = 0, \quad 0 \leq x \leq L$$

is

$$u_{\tilde{\alpha}}(x, y) = \sum_{n=1}^{\infty} \tilde{A}_n \sin\frac{n\pi y}{M} \sinh\frac{n\pi}{M}(L - x)$$

where

$$\tilde{A}_n \cosh\frac{n\pi L}{M} = \frac{2}{M}\int_0^M \tilde{\alpha}(y)\sin\frac{n\pi y}{M}dy$$

3. The problem is $u_{rr} + u_r/r + u_{\theta\theta}/r^2 = 0$, $0 < r < 1$, $|\theta| \leq \pi$; $u(1, \theta) = f(\theta)$. We also want to find the largest and smallest temperatures in the disk.

(a). Since $f(\theta) = 3\sin\theta$, we have from text formulas (10) and (11) that $A_0 = 0$,

$$A_n = \frac{1}{\pi}\int_{-\pi}^{\pi} 3\sin\theta\cos n\theta\, d\theta = 0 \qquad B_n = \frac{1}{\pi}\int_{-\pi}^{\pi} 3\sin\theta\sin n\theta\, d\theta = \begin{cases} 3, & n = 1 \\ 0, & n > 1 \end{cases}$$

The solution is $u(r, \theta) = 3r\sin\theta$. The maximum and minimum temperatures in the unit disk are ± 3, respectively.

(c). If $f(\theta) = 0$ for $-\pi < \theta < 0$, $f(\theta) = 1$ for $0 < \theta < \pi$, we have

$$A_n = \frac{1}{\pi}\int_{-\pi}^{\pi} f(\theta)\cos n\theta\, d\theta = \begin{cases} 1, & n = 0 \\ 0, & n > 0 \end{cases}$$

$$B_n = \frac{1}{\pi} \int_{-\pi}^{\pi} f(\theta) \sin n\theta \, d\theta = \frac{1 - (-1)^n}{n\pi} = \begin{cases} 2n/\pi, & n \text{ odd} \\ 0, & n \text{ even} \end{cases}$$

The solution is

$$u(r, \theta) = \frac{1}{2} + \frac{2}{\pi} \sum_{odd \ n} \frac{1}{n} r^n \sin n\theta$$

Since the highest and lowest temperatures must be reached on the perimeter, we have $\max u = \max f = 1$ and $\min u = \min f = 0$.

CHAPTER

11

Series Solutions: Bessel Functions, Legendre Polynomials

Chapters 3 and 4 give the general theory, solution techniques, and properties of solutions for a scalar linear second-order ODE, but with little said about how to find solution formulas if the coefficients are nonconstant. The solutions of nonconstant-coefficient second-order linear ODEs cannot usually be found in closed form, but can be written as power series or "Frobenius" series, and that is the topic of this chapter. It is this kind of second-order ODE that often arises when separating variables in second-order PDEs written in polar, cylindrical, or spherical coordinates, and we will see some of these PDEs in Sections 11.1, 11.7, and 11.8. That is the reason we have placed this chapter here. On the other hand, aside from the PDE examples of steady temperature models, this chapter could be covered immediately following Chapter 4, and that is when we often cover it in our courses. In Section 11.1 (and Appendix B.2) we review properties of power series; of particular importance are the techniques of reindexing and the Identity Theorem. The notion of an ordinary point for a second-order linear ODE is introduced in Section 11.2, and the techniques needed for constructing convergent power series solutions in the neighborhood of an ordinary point are introduced via examples. The two best known (and widely applied) ODEs with nonconstant coefficients are the Legendre and the Bessel ODEs; these are treated in Sections 11.3, 11.6, and 11.7. Regular singular points are defined in Section 11.4 and discussed mostly in the context of the Euler ODEs. In Sections 11.5, 11.6, and 11.7 the methods of Frobenius are introduced in order to construct series solutions near regular singular points. These methods are straightforward, but involve much algebraic manipulation; a CAS may be appropriate here. To provide an alternative application of series techniques, we model the behavior of an aging spring in Section 11.1 by a second-order linear ODE with nonconstant coefficients and then return to that model in Sections 11.2 and 11.7.

The material of this section is now "classical" in the sense that it was largely completed by the end of the 19th century. It still is widely used in the applications of ODEs, especially when working with those ODEs that arise when variables are separated in a PDE.

11.1 Aging Springs and Steady Temperatures

Comments

Series expansions are central in analysis and, for at least 150 years, in ODEs as well. In this section we take the first steps in developing series techniques for solving a second-order linear ODE in y whose coefficients are polynomials or convergent power series in the independent variable, expressing the solution as a series in powers of the independent variable (usually x in this chapter, but occasionally t as in the aging spring model). The reason for this choice of variables is mainly historical: those were the variables used when the PDEs for heat flow in a rod and the displacement of a vibrating string were originally reduced to ODEs by the technique of separating variables (Chapter 10).

---

1. Apply the Ratio Test (Appendix B.2, item 3) to each series to find the interval of convergence.

    (a). The series is $\sum x^n/n!$ and the ratio of successive terms is

    $$\left| \frac{x^{n+1}/(n+1)!}{x^n/n!} \right| = \frac{1}{n+1}|x| \to 0 \quad \text{as} \quad n \to \infty$$

    so the interval of convergence is $-\infty < x < \infty$.

    (c). In this case the series is $\sum x^{2n+1}/(2n+1)!$ and the ratio of successive terms is

    $$\left| \frac{x^{2n+3}/(2n+3)!}{x^{2n+1}/(2n+1)!} \right| = \frac{1}{(2n+3)(2n+2)}|x^2| \to 0 \quad \text{as} \quad n \to \infty$$

    so the convergence interval is $-\infty < x < \infty$.

3. In each case we have to lower or raise the power of x and (to compensate) raise or lower the bottom value of the range of the index by the same amount.

    (a). The power of x must be lowered one step, so the lower limit of the range must be raised one step: $\sum_{n=0}^{\infty} 2(n+1)x^{n+1}/n! = \sum_{n=1}^{\infty} 2nx^n/(n-1)!$.

    (c). The power of x must be lowered one unit and the lower range on the sum raised one unit: $\sum_{n=1}^{\infty}(-1)^{n-1}x^{n+1}/(n(n+1)) = \sum_{n=2}^{\infty}(-1)^n x^n/(n(n-1))$. Note that $(-1)^{n-2} = (-1)^n$, and we have selected the simpler form.

5. In each case we have $\sum_n [\cdot]x^n = 0$, so by the Identity Theorem (Appendix B.2, item 5), we must have $[\cdot] = 0$ for each n. This leads to the desired recursion formula.

    (a). By the Identity Theorem, $na_n - n + 2 = 0$ for $n = 1, 2, \ldots$, so $a_n = 1 - 2/n$.

    (c). In this case, we have $(n+2)(n+1)a_{n+2} = a_n$, $n = 0, 1, 2, \ldots$. So

    $$a_{n+2} = \frac{1}{(n+2)(n+1)}a_n = \frac{1}{(n+2)(n+1)}\left(\frac{1}{n(n-1)}a_{n-2}\right) = \cdots$$

    $$= \frac{1}{n+2} \cdot \frac{1}{n+1} \cdot \frac{1}{n} \cdot \frac{1}{n-1} \cdots \frac{1}{2}a_0$$

$$= \frac{1}{(n+2)!} a_0$$

if n is even. If n is odd, we get $a_{n+2} = a_1/(n+2)!$. In any case a_n is a_0 or a_1 divided by $n!$ if $n \geq 2$.

7. We want to use a series $y = \sum_0^\infty a_n x^n$ to solve the IVP, $y'' - 4y = 0$, $y(0) = 2$, $y'(0) = 2$.

(a). Let $y = \sum_{n=0}^\infty a_n x^n$. Then $y'' = \sum_{n=2}^\infty n(n-1)a_n x^{n-2}$, which is $\sum_{n=0}^\infty (n+2)(n+1)a_{n+2}x^n$ after reindexing. So $y'' - 4y = \sum_{n=0}^\infty [(n+2)(n+1)a_{n+2} - 4a_n]x^n = 0$. By the Identity Theorem, $(n+2)(n+1)a_{n+2} = 4a_n$, $n = 0, 1, 2, \ldots$. Since y has the form $y = a_0 + a_1 x + a_2 x^2 + \cdots$ and we require that $y(0) = 2$ and $y'(0) = 2$, we might as well set $a_0 = 2$, $a_1 = 2$ to begin with. Then from the recursion identity, $a_{n+2} = 4a_n/(n+2)(n+1)$, we see (much as in 5(c)), that $a_n = 2^n a_0/n!$ if n is even, and $2^{n-1}a_1/n!$ if n is odd, where $n > 1$. The solution is

$$y = \sum_{n=0}^\infty \frac{(2x)^{2n+1}}{(2n+1)!} + 2\sum_{n=0}^\infty \frac{(2x)^{2n}}{(2n)!} = \sum_{k=0}^\infty \frac{1}{k!}\left(\frac{3}{2} + \frac{1}{2}(-1)^k\right)(2x)^k$$

The exact solution is easily seen to be $y(x) = (3e^{2x} + e^{-2x})/2$, by the methods of Section 3.4. What we have found is the Taylor series for $(3e^{2x} + e^{-2x})/2$.

9. We want to find a_{100} if it is known that $\sum_0^\infty [(n+1)^2 a_{n+2} - n^2 a_{n+1} + (n-1)a_n]x^n = 0$. The recursion identity here is $(n+1)^2 a_{n+2} - n^2 a_{n+1} + (n-1)a_n = 0$, that is, $a_{n+2} = [n^2 a_{n+1} - (n-1)a_n]/(n+1)^2$, $n = 0, 1, 2, \ldots$, where $a_0 = a_1 = 1$. A computer may easily be programmed to calculate the coefficients using the identity and the initial data $a_0 = a_1 = 1$. In fact, $a_{100} \approx -1.0629 \times 10^{-6}$.

11. Group project.

11.2 Series Solutions Near an Ordinary Point

Comments

Ordinary points and singular points are defined for the ODE $y'' + P(x)y' + Q(x)y = 0$, and power series solutions are constructed in powers of $x - x_0$, where x_0 is an ordinary point. In all the examples and in the problems, x_0 is always taken to be zero for simplicity; in any case the variable change $z = x - x_0$ could be applied. Four of the famous "named" ODEs of classical analysis are introduced in the text and the problem set: the Legendre, Hermite (a particular case), Airy, and Mathieu ODEs. A series solution of the ODE for the aging spring is obtained and its usefulness examined. The Convergence Theorem for the series solutions about an ordinary point is presented, but not proved. In the text and in the problems when dealing with a specific power series we rely on the Ratio Test, rather than the Convergence Theorem, to determine intervals of convergence. Lots of interesting material here, but it may seem to be endless calculations with series. As in the first section, one algebraic mistake in working out the coefficients of a series solution, and the rest of the calculations are doomed. We think that every reader should carry through "by hand" the calculations of at least one of the series solutions, so learning the need for careful bookkeeping and methodical procedure, especially in reindexing and obtaining the recursion formula for the coefficients.

1. The definitions of ordinary point and singular point refer to the ODE $y'' + P(x)y' + Q(x)y = 0$.

   (a). The ODE is $y'' + k^2 y = 0$. Since $P(x) = 0$ and $Q(x) = k^2$ are real analytic at every point, every point x_0 is ordinary.

   (c). The ODE is $y'' + y'/x - y = 0$. Since $P(x) = 1/x$ is undefined at $x_0 = 0$, $P(x)$ is not real analytic at $x_0 = 0$, so 0 is a singular point.

3. For **(a)** and **(c)**, the functions $P(x)$ and $Q(x)$ are polynomials, so real analytic for all x. In every case the initial conditions are $y(0) = 1$, $y'(0) = 1$.

   (a). The ODE is $y'' + 4y = 0$. Here the most direct way is to note that the general solution is

   $$y = C_1 \cos 2x + C_2 \sin 2x$$

   $$= C_1 \sum_{k=0}^{\infty} [(-1)^k/(2k)!](2x)^{2k} + C_2 \sum_{k=0}^{\infty} [(-1)^k/(2k+1)!](2x)^{2k+1}$$

   where we have used the Taylor series for $\cos 2x$ and $\sin 2x$ about $x_0 = 0$. By the Ratio Test, for the cosine series,

   $$\lim_{n\to\infty} \left| \frac{(2x)^{2n+2}(2n)!}{(2x)^{2n}(2n+2)!} \right| = \lim_{n\to\infty} \left| \frac{(2x)^2}{(2n+2)(2n+1)} \right| = 0$$

   The series converges for all x. The Ratio Test gives the same result for the series for $\sin(2x)$. The solution of the IVP is $y = \cos x + (\sin x)/2$. See Fig. 3(a).

   (c). The ODE is $y'' + xy' + 3y = 0$. Let the general solution be $y = \sum_{n=0}^{\infty} a_n x^n$, so $y' = \sum_{n=1}^{\infty} n a_n x^{n-1}$, $y'' = \sum_{n=2}^{\infty} n(n-1)a_n x^{n-2}$. Then $y'' + xy' + 3y = 0$ becomes

   $$\sum_{n=2}^{\infty} n(n-1)a_n x^{n-2} + \sum_{n=1}^{\infty} n a_n x^n + \sum_{n=0}^{\infty} 3a_n x^n = 0$$

   In order to include all three sums in a single summation, we must reindex the first sum and lower the range on the second from 1 to 0 (easy to do since $n = 0$ only gives the term 0 in the second sum):

   $$\sum_{n=0}^{\infty} [(n+2)(n+1)a_{n+2} + n a_n + 3a_n]x^n = 0$$

   By the Identity Theorem, $a_{n+2} = -(n+3)a_n/[(n+2)(n+1)]$. If n is even [odd], we can push this recursion formula for the coefficients back to a_0 [a_1]. This is most easily done by considering the two cases separately, first n even ($n = 2k$) and then n odd ($n = 2k + 1$).

   $$a_{2k} = -\frac{2k+1}{(2k)(2k-1)}a_{2k-2} = \left(-\frac{2k+1}{(2k)(2k-1)}\right)\left(-\frac{2k-1}{(2k-2)(2k-3)}a_{2k-4}\right)$$

   $$= \cdots = \frac{(-1)^k(2k+1)(2k-1)\cdots 3}{(2k)!}a_0$$

Since

$$(2k+1)(2k-1)\cdots 3 = \frac{(2k+1)(2k)(2k-1)(2k-2)\cdots 3\cdot 2}{(2k)(2k-2)\cdots 2} = \frac{(2k+1)!}{2^k k!}$$

we have that

$$a_{2k} = (-1)^k \frac{(2k+1)!}{(2k)!2^k k!} = (-1)^k \frac{2k+1}{2^k k!} a_0$$

In exactly the same way, it may be shown that

$$a_{2k+1} = -\frac{2k+2}{(2k+1)(2k)}a_{2k-1} = \cdots = (-1)^k \frac{(2k+2)(2k)\cdots 4}{(2k+1)!}a_1 = (-1)^k \frac{2^k(k+1)!}{(2k+1)!}a_1$$

We may write the general solution as

$$y = a_0 \sum_{k=0}^{\infty}(-1)^k \frac{(2k+1)}{2^k k!}x^{2k} + a_1 \sum_{k=0}^{\infty}(-1)^k \frac{2^k(k+1)!}{(2k+1)!}x^{2k+1}$$

where a_0 and a_1 are arbitrary constants. We may test the convergence of the series by the Ratio Test. For the first series we have that the magnitude of the ratio of successive terms is $|(2k+3)x^2/[2(k+1)(2k+1)]|$, which $\to 0$ as $k \to \infty$. The series converges for all x. It may be shown in much the same way that the second series also converges for all x. Since $y(0) = 1$ and $y'(0) = 1$, we set $a_0 = a_1 = 1$ to get the solution that meets the initial conditions. See Fig. 3(c).

Problem 3(a).

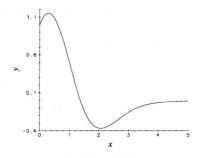

Problem 3(c).

5.   The Matthieu equation is $y'' + (a+b\cos\omega x)y = 0$. Set $a = 1$, $b = 2$, $\omega = 1$, and $y(0) = 1$, $y'(0) = 0$. Since we only want the first few terms, we shall not attempt to find the general coefficient of the series solution. The initial data imply that $a_0 = 1$ and $a_1 = 0$ if $y = a_0 + a_1 x + a_2 x^2 + \cdots$. Since $\cos x$ has Taylor series $1 - x^2/2! + x^4/4! - x^6/6! + \cdots$ about $x_0 = 0$, we see that inserting the series for y and for $\cos x$ in the ODE and using item 6(ii) of Appendix B.2 to determine the first few terms of the product series for $(2\cos x)y(x)$, we have that

$$0 = y'' + (1+2\cos x)y = (3+2a_2) + 6a_3 x + (12a_4 + 3a_2 - 1)x^2 + (3a_3 + 20a_5)x^3$$

$$+ (30a_6 + 3a_4 - a_2 + \frac{1}{12})x^4 + \cdots$$

So $a_2 = -3/2$, $a_3 = 0$, $a_4 = 13/48$, $a_5 = 0$, $a_6 = -23/288$, and the series for the solution is

$$y = 1 - \frac{3}{2}x^2 + \frac{13}{48}x^4 - \frac{23}{288}x^6 + \cdots$$

7. The ODE is $y'' + (1 - e^{-x^2})y = 0$. Let $y = a_0 + a_1 x + a_2 x^2 + \cdots$. Since the Taylor series expansion of e^{-x^2} about $x_0 = 0$ is $1 - x^2 + x^4/2! - x^6/3! + \cdots$, we have that

$$y'' + y - e^{-x^2}y = 2a_2 + 6a_3 x + (12a_4 + a_0)x^2$$
$$+ (20a_5 + a_1)x^3 + (30a_6 + a_2 - \frac{a_0}{2})x^4 + \cdots$$

where we have used Appendix B.2, item 6(ii) to calculate the first few terms of the product series for $(1 - e^{-x^2})y(x)$. The coefficients a_0 and a_1 are arbitrary, while $a_2 = a_3 = 0$, $a_4 = -a_0/12$, $a_5 = -a_1/20$, $a_6 = a_0/60$. We have the general solution

$$y = a_0(1 - \frac{1}{12}x^4 + \frac{1}{60}x^6 + \cdots) + a_1(x - \frac{1}{20}x^5 + \cdots)$$

11.3 Legendre Polynomials

Comments

The Legendre polynomials $\{P_n : n = 0, 1, 2, \ldots\}$ are the best known of the sets of orthogonal polynomials, and so we have chosen to give them and their ODEs a section of their own. This section, its problem set, and the full solutions of the problems should be regarded as an abbreviated compendium of properties of Legendre polynomials, other solutions of the Legendre ODEs, and other sets of orthogonal functions.

The Hermite ODEs and polynomials H_n and the Chebyshev ODEs and polynomials T_n make their appearance in the problem set. The reason the Chebyshev polynomials are labeled T_n and not C_n illustrates the difficulty in using the Latin alphabet to transliterate Cyrillic letters (Chebyshev was a well-known 19th century Russian mathematician). Sometimes one sees Tschebysheff, or Tchebyshev, or Chebyshov. Philip J. Davis has written an amusing and mathematically informative book based on his difficulties with these various transliterations [*The Thread: A Mathematical Gem*; Boston, Birkhaüser, 1983].

1. Assume that $u(x)$ is a solution of the ODE $y'' + a(x)y' + b(x)y = 0$ and that $u(x)$ is not the trivial solution $y = 0$, all x. Then a second solution $v(x)$, $|x| < 1$, independent of $u(x)$ can be found by solving the ODE

$$W[u, v](x) = uv' - u'v = e^{-\int^x a(s)\,ds}$$

In Legendre's ODEs of order 0 and 1, we have that $a(x) = -2x/(1 - x^2)$, $u_0(x) = 1$, and

$u_1(x) = x$, respectively. So $v_0' = e^{\int^x 2s/(1-s^2)ds} = C_1/(1-x^2)$, and (using an integral table)

$$v_0 = \frac{C_1}{2} \ln \frac{1+x}{1-x} + C_2, \quad (C_1 > 0)$$

Also, $xv_1' - v_1 = C_1/(1-x^2)$, $v_1' - v_1/x = C_1/x(1-x^2)$.

$$v_1 = e^{\int^x (1/s)\,ds}[C_3 + \int^x e^{-\int^s (1/\tau)\,d\tau}\frac{C_1}{s(1-s^2)}\,ds]$$

$$= x[C_3 + C_1 \int^x \frac{C_1}{s^2(1-s^2)}\,ds]$$

$$= x[C_3 + C_1(-\frac{1}{x} + \frac{1}{2} \ln \frac{1+x}{1-x})] \quad (C_1 > 0)$$

where we have used a table of integrals. So, second independent solutions for Legendre's equations of order 0 and 1 on the interval $|x| < 1$ are, respectively,

$$v_0 = \frac{C_1}{2} \ln \frac{1+x}{1-x} + C_2 \quad \text{and} \quad v_1 = x[C_3 + C_1(-\frac{1}{x} + \frac{1}{2} \ln \frac{1+x}{1-x})]$$

where C_1, C_2 and C_3 are constants and $C_1 > 0$. See Graph 1 and Graph 2, respectively, for graphs of v_0 and v_1, where $C_1 = 1$, $C_2 = 0$, $C_3 = 0$.

Problem 1, Graph 1.

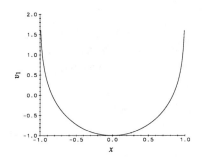

Problem 1, Graph 2.

3. Suppose that $Q(x)$ is any polynomial of degree $k < n$. We want to show that the integral $\int_{-1}^{1} P_n(x)Q(x)\,dx = 0$. We prove by induction that there are constants $c_0, \ldots, c_k$ such that $Q(x) = \sum_{j=0}^{k} c_k P_k(x)$. This is certainly the case for $k = 0$ since $Q(x) = Q(0) \cdot 1 = Q(0) \cdot P_0(x)$. Suppppose the result is true for all polynomials Q of degree $\leq k$ (the induction hypothesis). Let Q be a polynomial of degree $k + 1$. Let $Q = a_0 + \cdots + a_{k+1}x^{k+1}$ and $P_{k+1} = b_0 + \cdots + b_{k+1}x^{k+1}$. Then $Q(x) = a_{k+1}P_{k+1}(x)b_{k+1}^{-1} + R(x)$, where $R(x)$ is a polynomial of degree k or less. $R(x) = \sum_{j=0}^{k} \alpha_j P_j(x)$ for some constants α_j by the induction hypothesis, so $Q(x)$ is a linear combination of $P_0(x), \ldots, P_{k+1}(x)$ and we are done. Note that the induction proof did not use the specific form of the Legendre polynomials, only that $P_n(x)$ is a polynomial of degree n.

Now to the problem itself: suppose $Q(x)$ is any polynomial of degree of $k < n$. Then

$Q = c_0 P_0 + \cdots + c_k P_k$ and

$$\int_{-1}^{1} P_n(x) Q(x)\, dx = \sum_{j=0}^{k} c_j \int_{-1}^{1} P_n(x) P_j(x)\, dx = 0$$

by Theorem 11.3.2.

5. We work with a family $\{R_n(x)\}$ of polynomials indexed by degree and we assume that for all $n, m, n \neq m$ we have

$$\int_{a}^{b} R_n(x) R_m(x) \rho(x)\, dx = 0$$

We say that $\{R_n : n = 0, 1, 2, \ldots\}$ is an orthogonal family with respect to the density ρ.

(a). Theorem 11.3.2 states that

$$< P_n, P_m > = \int_{-1}^{1} P_n(x) P_m(x)\, dx = \begin{cases} 0, & n \neq m \\ 2/(2n+1), & n = m \end{cases}$$

So, $a = -1$ and $b = 1$, $\rho(x) = 1$ for this orthogonal family.

(c). From formula (10) for Legendre polynomials and their recursion formula, we see that (10) has the form $R_n = (a_n x + b_n) R_{n-1} + c_n R_{n-2}$ if $a_n = 2 - 1/n$, $b_n = 0$, $c_n = -(1 - 1/n)$.

7. (a). Chebyshev's equation is $(1 - x^2) y'' - x y' + n^2 y = 0$. Let $y = \sum_{k=0}^{\infty} a_k x^k$ and proceed much as in the preceding problems to write the ODE in the form $\sum_{k=0}^{\infty} [\cdots] x^k = 0$. In this case equating $[\cdots]$ to 0 gives the recursion formula,

$$a_{k+2} = \frac{k^2 - n^2}{(k+1)(k+2)} a_k$$

which can then be used to find the general solution

$$y = a_0 \left[1 - \frac{n^2}{2!} x^2 - \frac{(4 - n^2) n^2}{4!} x^4 - \frac{(16 - n^2)(4 - n^2) n^2}{6!} x^6 - \cdots \right]$$
$$+ a_1 \left[x + \frac{1 - n^2}{3!} x^3 + \frac{(9 - n^2)(1 - n^2)}{5!} x^5 + \cdots \right].$$

If n is an even [odd] integer, the first [second] series in brackets terminates and polynomial solutions are obtained. The polynomials are even functions if n is even, odd functions if n is odd. Let $T_0 = 1$ be the zeroth Chebyshev polynomial, and let the n-th Chebyshev polynomial $T_n(x)$ be the polynomial solution with leading coefficient 2^{n-1}.

(c). Use the given recursion formula $T_n = 2x T_{n-1} - T_{n-2}$, $T_0 = 1$, $T_1 = x$; therefore, $T_2 = 2x^2 - 1$, $T_3 = 4x^3 - 3x$, $T_4 = 8x^4 - 8x^2 + 1$, $T_5 = 16x^5 - 20x^3 + 5x$. See Fig. 7 for the graphs of T_0 and T_4 (solid), T_1 and T_5 (long dashes), T_2 (short dashes), and T_3 (long/short dashes). The graphs suggest that $T_n(x)$ has n simple real roots in $(-1, 1)$ since each T_n seems to cross the x-axis n times (each crossing point is a root).

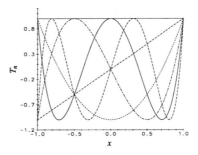

Problem 7(c).

Background Material: Proof of Bounds for Legendre Polynomial

Squaring the identity $P'_n = n P_{n-1} + x P'_{n-1}$, multiplying through by $(1 - x^2)/n^2$, and adding it to the square of the identity $P_n = x P_{n-1} + [(x^2 - 1)/n] P'_{n-1}$, we obtain, for $|x| \leq 1$,

$$0 \leq \frac{1-x^2}{n^2}(P'_n)^2 + P_n^2 = P_{n-1}^2 + \frac{1-x^2}{n^2}(P'_{n-1})^2 < P_{n-1}^2 + \frac{1-x^2}{(n-1)^2}(P'_{n-1})^2 \qquad \text{(i)}$$

If we have (for $n = 1$) an upper bound for the right side of the inequalities in (i), that bound provides an upper bound for the second term in (i) for all $n \geq 1$. But the right side of (i) for $n = 1$ and for any x, $|x| \leq 1$, has value 1 since $P_0 = 1$ and $P'_0 = 0$. From all this, we conclude that

$$0 \leq \frac{1-x^2}{n^2}(P'_n(x))^2 + P_n^2(x) \leq 1, \qquad |x| < 1$$

so, $|P_n(x)| \leq 1$ for $|x| \leq 1$.

11.4 Regular Singular Points

Comments

Now that we know how to construct power series solutions of a linear ODE in powers of $x - x_0$, where x_0 is an ordinary point, it is time to address the challenging problem of constructing series solutions if x_0 is singular. That can be done if x_0 is "regular singular" (isn't that an oxymoron?), and we start out by defining just what a regular singular point is. The simplest class of ODEs with regular singular points are the Euler ODEs, and most of the section is devoted to the construction of their solutions. The nice thing about an Euler ODE is that the intricate series techniques of the preceding three sections are not needed. Students (and the instructor) will have a welcome respite from the algebraic rigors of dealing with series. The indicial polynomial appears here, and should be emphasized because it plays a critical role in the analysis of the remaining sections of the chapter. There are some interesting pictures of solution behavior of an Euler ODE near its regular singularity at the origin. It should be noted that the basic Existence and Uniqueness Theorem (Theorem 3.2.1) doesn't work at the singularity, and an IVP may have no solution at all or else infinitely many (see Example 11.4.6).

1. The generic ODE here is $a_2 y'' + a_1 y' + a_0 y = 0$, which in normal form is $y'' + (a_1/a_2)y' + (a_0/a_2)y = 0$. The tests for a regular or irregular singularity at $x = x_0$ are concerned with the functions $(x - x_0)(a_1/a_2)$ and $(x - x_0)^2(a_0/a_2)$.

(a). The ODE is $x^2 y'' + xy' + y = 0$. Here, $a_2 = x^2$, $a_1 = x$, $a_0 = 1$, while $x_0 = 0$. We have

$$\lim_{x \to 0} \frac{xa_1}{a_2} = \lim_{x \to 0} \frac{x^2}{x^2} = 1, \quad \lim_{x \to 0} \frac{x^2 a_0}{a_2} = \lim_{x \to 0} \frac{x^2}{x^2} = 1$$

and 0 is a regular singular point for $x^2 y'' + xy' + y = 0$.

(c). The ODE is $x(1 - x)y'' + (1 - 2x)y' - 4y = 0$. In this problem $x_0 = 1$ is the singular point to be tested. We have $a_2 = x(1 - x)$, $a_1 = (1 - 2x)$, $a_0 = -4$. We have

$$\lim_{x \to 1} \frac{(x - 1)(1 - 2x)}{x(1 - x)} = 1, \quad \lim_{x \to 1} \frac{(x - 1)^2(-4)}{x(1 - x)} = \lim_{x \to 1} \frac{4(x - 1)}{x} = 0$$

and $x_0 = 1$ is a regular singular point.

3. Each of the equations may be written as an Euler equation $x^2 y'' + p_0 xy' + q_0 y = 0$, and the solutions are determined by the roots r_1 and r_2 of the indicial polynomial $r^2 + (p_0 - 1)r + q_0$. The three types of solutions are given in (13) in the text. In each case we list the indicial polynomial, its roots, and then the general solution; c_1 and c_2 denote arbitrary constants. It is assumed throughout that $x > 0$, and absolute value signs around x are not needed. See the appropriate figures for plots of some solutions of each equation. For negative x, replace x by $|x|$ in all formulas.

(a). $r^2 - r - 6$; $r_1 = -2$, $r_2 = 3$; $y = c_1 x^{-2} + c_2 x^3$.

(c). $r^2 + 9$; $r_1 = 3i$, $r_2 = -3i$; $y = c_1 \cos(3 \ln x) + c_2 \sin(3 \ln x)$.

(e). $r^2 - 2r + 5$; $r_1 = 1 + 2i = \bar{r}_2$; $y = x[c_1 \cos(2 \ln x) + c_2 \sin(2 \ln x)]$.

Problem 3(a).

Problem 3(c).

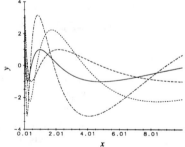

Problem 3(e).

5. The solutions of $x^2 y'' + p_0 xy' + q_0 y = 0$ are given by (13), and we see that every solution tends to 0 as $x \to +\infty$ if and only if r_1 and r_2 are negative or else have negative real parts. This is clear if r_1 and r_2 are real and distinct. If $r_1 = r_2 < 0$, then the term $x^{r_1} \ln x = \ln(x)/x^{-r_1} \to 0$ as $x \to +\infty$ by L'Hôpital's Rule. If $r_1 = a + bi = \bar{r}_2$ and $a < 0$, $b \neq 0$, then $|x^a \cos(b \ln x)| \le |x|^a \to 0$ as $x \to +\infty$, and similarly for $x^a \sin(b \ln x)$. The roots r_1 and r_2 of the indicial polynomial $r^2 + (p_0 - 1)r + q_0$ have negative real parts if and only if $p_0 > 1$ and $q_0 > 0$, as can be seen by using the quadratic formula for the roots.

11.5 Series Solutions Near Regular Singular Points, I

Comments

In the next three sections of the chapter we take up the techniques for constructing a pair of independent solutions of a homogeneous second-order linear ODE with nonconstant coefficients near a regular singularity. In Section 11.5 we address the simplest case when the roots of the indicial polynomial do *not* differ by an integer. We also show that even if the roots do differ by an integer we can always construct a solution corresponding to the larger root. The process by which Frobenius series expansions of solutions are constructed is similar to the methods used in Section 11.2 to find power series expansions of solutions near an ordinary point, but with the twist that the powers of the independent variable need not be integers. As before, the successful construction of the series requires patience, algebraic "smarts," and careful bookkeeping. Once again, a CAS may be helpful. We have noticed over the years that readers have a lot of trouble with this material. But those who finally "see" how to do it are quite proud of their accomplishment.

1. The detailed solution process for finding y_1 and y_2 is given only for **(a)**. In the other cases only the indicial polynomial, its roots r_1 and r_2, the recursion formula and the corresponding independent solutions, $y_1 = x^{r_1} \sum_{n=0}^{\infty} a_n(r_1)x^n$ and $y_2 = x^{r_2} \sum_{n=0}^{\infty} a_n(r_2)x^n$, are given. In each case $r_1 - r_2$ is not an integer, and the Method of Forbenius given in this section produces the two independent solutions y_1 and y_2.

(a). The ODE is $9x^2 y'' + 3x(x+3)y' - (4x+1)y = 0$. The equation in standard form is $x^2 y'' + x(1 + x/3)y' + (-1/9 - 4x/9)y = 0$. So $p_0 = 1$ and $q_0 = -1/9$. The indicial polynomial is $r^2 + (p_0 - 1)r + q_0 = r^2 - 1/9$, and its roots are $r_1 = 1/3$, $r_2 = -1/3$. Let $y = \sum_{n=0}^{\infty} a_n x^{n+r}$ be a solution. Then

$$y' = \sum_{n=0}^{\infty}(n+r)a_n x^{n+r-1}, \quad y'' = \sum_{n=0}^{\infty}(n+r)(n+r-1)a_n x^{n+r-2}$$

Inserting these series into the ODE,

$$9x^2 y'' + 3x(x+3)y' - (4x+1)y = 0$$

reindexing, and rearranging, we have that

$$\sum_{n=0}^{\infty}[9(n+r)(n+r-1) + 9(n+r) - 1]a_n x^{n+r} + \sum_{n=1}^{\infty}[3(n+r-1) - 4]a_{n-1}x^{n+r} = 0$$

Note that the term corresponding to $n = 0$ in the first summation above is zero because $r^2 = 1/9$. The index in that sum actually runs from $n = 1$ to ∞, and the two summations can be combined because they have the same index range and involve the same power of x. Applying the Identity Theorem to the coefficient of x^{n+r}, we obtain the recursion relation:

$$[9(n+r)(n+r-1) + 9(n+r) - 1]a_n = -[3(n+r-1) - 4]a_{n-1}$$

and, solving this relation for a_n,

$$a_n = -\frac{3(n+r-1) - 4}{9(n+r)^2 - 1}a_{n-1}$$

So for $r = 1/3$,

$$a_n = -(n-2)a_{n-1}/[n(3n+2)], \quad n \geq 1$$

Setting $a_0 = 1$, we have that $a_1 = 1/5$, while $a_n = 0$, $n \geq 2$. We have a solution corresponding to $r = 1/3$:

$$y_1 = x^{1/3} + \frac{1}{5}x^{4/3}$$

For $r = -1/3$,

$$a_n = -(3n-8)a_{n-1}/[3n(3n-2)]$$
$$= a_0(-1)^n(3n-8)\cdots(-2)(-5)/[3^n n!(3n-2)(3n-5)\cdots 1]$$
$$= (-1)^n\frac{10}{3^n n!(3n-2)(3n-5)}a_0, \quad n \geq 2$$

So a second independent solution (corresponding to $r = -1/3$) is

$$y_2 = x^{-1/3}\sum_{n=0}^{\infty} a_n x^n$$

where $a_0 = 1$ and a_n for $r = -1/3$ is given above. The series converge for all x but the general solution $y = c_1 y_1 + c_2 y_2$ is not defined at $x = 0$, because of the factor $x^{-1/3}$ in y_2.

(c). The ODE is $x^2 y'' + x(1-x)y' - 2y = 0$. The indicial polynomial is $r^2 - 2$ and its roots are $r_1 = \sqrt{2}$, $r_2 = -\sqrt{2}$. The recursion formula is

$$a_n(r) = (n+r-1)a_{n-1}/[(n+r)^2 - 2]$$

So

$$a_n(\pm\sqrt{2}) = \frac{(n \pm \sqrt{2} - 1)(n \pm \sqrt{2} - 2)\cdots(\pm\sqrt{2})a_0}{n!(n \pm 2\sqrt{2})(n - 1 \pm 2\sqrt{2})\cdots(1 \pm 2\sqrt{2})}$$

A pair of independent solutions is

$$y_1 = x^{\sqrt{2}}\sum_{n=0}^{\infty} a_n(\sqrt{2})x^n \quad \text{and} \quad y_2 = x^{-\sqrt{2}}\sum_{n=0}^{\infty} a_n(-\sqrt{2})x^n$$

where $a_0 = 1$ and $a_n(\pm\sqrt{2})$ is given above.

3. The ODE is $3x^2 y'' + 5xy' - e^x y = 0$. The Taylor series for e^x about $x_0 = 0$ is $1 + x + x^2/2! + x^3/3! + \cdots$. So $p_0 = 5/3$, $q_0 = -1/3$ for the ODE in nonstandard form

$$x^2 y'' + 5xy'/3 - e^x y/3 = x^2 y'' + 5xy'/3 + (-1/3 - x/3 - \cdots)y = 0$$

The indicial polynomial is $r^2 + 2r/3 - 1/3$ and $r_1 = 1/3$, $r_2 = -1$. There are independent solutions of the form

$$y_1 = x^{1/3} + a_1 x^{4/3} + a_2 x^{7/3} + \cdots, \quad y_2 = x^{-1} + b_1 + b_2 x + \cdots$$

where we have set $a_0 = b_0 = 1$ in each series. Although we could use Frobenius Theorem I to determine the coefficients, it is probably simpler to make a direct substitution of y_1 and then y_2 into the ODE. For y_1 we have that

$$3x^2 y_1'' + 5xy_1' - e^x y_1 = (7a_1 - 1)x^{4/3} + (20a_2 - a_1 - 1/2)x^{7/3}$$
$$+ (39a_3 - a_2 - a_1/2 - 1/6)x^{10/3} + \cdots = 0$$

So $a_1 = 1/7$, $a_2 = 9/280$, $a_3 = 227/32760$ and

$$y_1 = x^{1/3} + x^{4/3}/7 + 9x^{7/3}/280 + 227x^{10/3}/32760 + \cdots$$

In the same way we see that

$$y_2 = x^{-1} - 1 - x/8 - 11x^2/360 + \cdots$$

5. Suppose $x^3 y'' + y = 0$ has a solution $y = \sum_{n=0}^{\infty} a_n x^{n+r}$ for some constant r and constants a_n, where the series $\sum_{n=0}^{\infty} a_n x^n$ converges in a nontrivial interval I containing 0. Then we obtain a power series after a certain amount of reindexing

$$x^3 y'' + y = a_0 x^r + \sum_{n=1}^{\infty} [(n+r-1)(n+r-2)a_{n-1} + a_n]x^{n+r} = 0$$

So $a_0 = 0$, and

$$a_n = -(n+r-1)(n+r-2)a_{n-1}$$

This implies that $a_1 = a_2 = \cdots = a_n = \cdots = 0$, and the only solution in power series form is the trivial solution ($y = 0$).

7. **(a).** The Legendre ODE of order p is $(1 - x^2)y'' - 2xy' + p(p+1)y = 0$. First, we let $x = s + 1$ so that the regular singularity at $x = 1$ is translated to the origin. Legendre's ODE for $z(s) = y(x)$ then becomes

$$[1 - (s+1)^2]\frac{d^2 z}{ds^2} - 2(s+1)\frac{dz}{ds} + p(p+1)z = 0$$

We need to write the ODE in the standard form $s^2 y'' + sp(s)z' + q(s)z = 0$. Since the coefficient of z'' is $1 - (s+1)^2 = s(-s-2)$, we multiply the ODE by s and divide by $-s - 2$ to obtain

$$s^2 \frac{d^2 z}{ds^2} + \frac{2(s+1)s}{s+2}\frac{dz}{ds} + \frac{p(p+1)(-s)}{s+2}z = 0$$

Since the Taylor series for $2(s+1)/(s+2)$ and $p(p+1)(-s)/(s+2)$ are, respectively, $1 + (\cdot)s + \cdots$, and $0 - p(p+1)s/2 + \cdots$, we see that $p_0 = 1$, $q_0 = 0$, the indicial polynomial is r^2, and the indicial roots are $r_1 = r_2 = 0$.

11.6   Bessel Functions

Comments

The classic examples of ODEs with a regular singularity are the Bessel equations, and this section and the next are mostly about these equations and their solutions. Along the way we define and give a few of the properties of another classic function, Euler's gamma function (or the interpolated factorial). The construction of the Bessel functions of the first kind follows the Frobenius steps outlined in Section 11.5. The gamma function has unusual properties and is fun to work with. Bessel functions of the first kind resemble decaying sinusoids in many ways. For later use in solving PDEs in Section 11.8, we discuss the positive zeros λ_n of Bessel functions J_p, the resemblance between $J_p(\lambda_n x)$ and $\sin(n\pi x)$, and the orthogonality properties.

1. Using (6), we have that

$$J_3(x) = \left(\frac{x}{2}\right)^3 \left[\frac{1}{3!} - \frac{1}{4!}\left(\frac{x}{2}\right)^2 + \frac{1}{2!5!}\left(\frac{x}{2}\right)^4 - \frac{1}{3!6!}\left(\frac{x}{2}\right)^6 + \cdots\right]$$

$$= x^3/48 - x^5/768 + x^7/30720 - x^9/2211840 + \cdots$$

3. According to Problem 6, Section 3.7, a second independent solution y_2 of Bessel's equation of order p satisfies the first-order linear equation $J_p y_2' - J_p' y_2 = x^{-1}$, where the term on the right side is $\exp[-\int^x a(s)\,ds]$ and $a(x) = x^{-1}$ [the normalized form $y'' + ay' + by = 0$ of Bessel's equation is $y'' + x^{-1}y' + (1 - p^2 x^{-2})y = 0$]. So $y_2' - J_p' y_2/J_p = 1/(xJ_p)$ and the integrating factor of the first-order linear ODE is $1/J_p$. We have $(y_2/J_p)' = 1/(xJ_p^2)$ and we can take $y_2 = J_p(x)\int^x (sJ_p^2(s))^{-1}\,ds$. The general solution of Bessel's equation of order p is $y = c_1 J_p(x) + c_2 J_p(x)\int^x (sJ_p^2(s))^{-1}\,ds$, c_1 and c_2 any constants.

5. See Figs. 5 for plots of J_0, J_1, J_2, J_3 (Graph 1) and $J_{1/3}$, $J_{4/3}$, $J_{7/3}$ (Graph 2).

Problem 5, Graph 1.

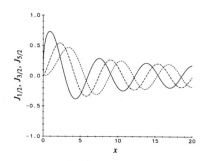

Problem 5, Graph 2.

7.   **(a).** From (11), we have $\Gamma(1/2) = \int_0^\infty x^{-1/2} e^{-x} \, dx$. Let's show that $\Gamma(1/2) = \sqrt{\pi}$. Let $x = u^2$. Then $\Gamma(1/2) = \int_0^\infty 2e^{-u^2} \, du$. To evaluate this integral, let $I = \int_0^\infty e^{-u^2} \, du$. Then

$$I^2 = \int_0^\infty e^{-u^2} \, du \int_0^\infty e^{-v^2} \, dv = \int_0^\infty \left(\int_0^\infty e^{-(u^2+v^2)} \, du \right) dv$$

$$= \int_0^{\pi/2} \int_0^\infty \left(re^{-r^2} \, dr \right) d\theta = -\frac{1}{2} \int_0^{\pi/2} \{e^{-r^2}\} \Big|_0^\infty \, d\theta = \pi/4$$

where we have introduced polar coordinates in the first quadrant of the uv plane to carry out the indicated double integral. So $I^2 = \pi/4$, $I = \sqrt{\pi}/2$, and $\Gamma(1/2) = 2I = \sqrt{\pi}$.

   (c). We have

$$\Gamma(3/2) = \Gamma(1/2 + 1) = \Gamma(1/2)/2 = \sqrt{\pi}/2$$

and

$$\Gamma(5/2) = \Gamma(3/2 + 1) = 3\Gamma(3/2)/2 = 3\sqrt{\pi}/4$$

Similarly,

$$\Gamma(-3/2) = \Gamma(-3/2 + 1)/(-3/2) = -2\Gamma(-1/2)/3 = 4\sqrt{\pi}/3$$

and

$$\Gamma(-5/2) = \Gamma(-5/2 + 1)/(-5/2) = -2\Gamma(-3/2)/5 = -8\sqrt{\pi}/15$$

9.   **(a).** Set $p = 0$ in equation (19) and obtain that $J_0' = -J_1$. So $\int^x J_1(t) \, dt = -J_0(x) + C$. Equation (18) for $p = 1$ yields that $(xJ_1)' = xJ_0$, and this becomes $\int^x tJ_0(t) \, dt = xJ_1(x) + C$.

   (c). $\int^x t^2 J_0 \, dt = \int^x t(tJ_0) \, dt$ [using **(a)**] $= \int^x t(tJ_1)' \, dt$ [integration by parts] $= x(xJ_1) - \int^x tJ_1 \, dt = x^2 J_1 - \int^x t(-J_0)' \, dt = x^2 J_1 + xJ_0 - \int^x J_0 \, dt$.

11. See the figure for the graph of $J_1(x)$, $0 \leq x \leq 30$. From the graph the 8th and 9th positive zeros λ_8, λ_9 appear to be about 25.9 and 29, respectively. The difference is slightly less than π.

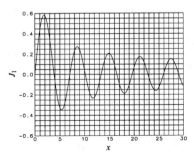

Problem 11.

Background Material: Verification of Bessel Recursion Relations

Let's verify formula (18). Multiply each side of (15) by x^p, differentiate term-by-term, and use the fact that $\Gamma(k + p + 1) = \Gamma(k + p)(k + p)$:

$$[x^p J_p]' = \frac{d}{dx} \left\{ \sum_{k=0}^{\infty} \frac{(-1)^k 2^p}{k! \Gamma(k + p + 1)} \left(\frac{x}{2}\right)^{2k+2p} \right\}$$

$$= \sum_{k=0}^{\infty} \frac{(-1)^k (k + p) 2^p}{k! \Gamma(k + p + 1)} \left(\frac{x}{2}\right)^{2k+2p-1}$$

$$= x^p \left(\frac{x}{2}\right)^{p-1} \sum_{k=0}^{\infty} \frac{(-1)^k}{k! \Gamma(k + p)} \left(\frac{x}{2}\right)^{2k} = x^p J_{p-1}$$

The verification of (19) is similar. To verify (16), multiply each side of (18) by x^{-p}, each side of (19) by x^p, and subtract the two:

$$x^{-p}[x^p J_p]' - x^p[x^{-p} J_p]' = J_{p-1} + J_{p+1}$$

$$x^{-p}[px^{p-1} J_p + x^p J_p'] - x^p[-px^{-p-1} J_p + x^{-p} J_p'] = J_{p-1} + J_{p+1}$$

from which (16) follows after a little algebraic rearrangement. Formula (17) is derived similarly. Note that (16)–(19) also hold for the functions J_{-p} if p is positive and not an integer.

11.7 Series Solutions Near Regular Singular Points, II

Comments

The cases where the roots of the indicial polynomial are equal or differ by an integer are addressed in this section. Two examples are worked out in detail so that the reader can see the step-by-step process for obtaining a second solution that involves a logarithm, once the first solution has been obtained by the methods of Section 11.5. Alternatively, the Wronskian Reduction Method outlined in Problem 6 of Section 3.7 also leads to a logarithmic term in the second solution. Bessel functions Y_p of the second kind are introduced and (briefly) discussed. The important point here is that $Y_p(x) \to -\infty$ as $x \to 0^+$. Finally, applications of all of this are made to the two problems introduced at the start of the chapter: the model of the aging spring, and the ODE arising when the PDE boundary problem model for steady temperatures in a cylinder has its variables separated. Both involve ODEs that reduce to a Bessel equation when new variables are introduced. This, incidentally, is another reason for the importance of the Bessel equations: many seemingly intractable ODEs can be reduced to a Bessel equation by a variable change.

1. It is straightforward to show that 0 is a regular singular point of each ODE; the details are omitted.

    (a). The ODE is $xy'' + (1 + x)y' + y = 0$. The equation in standard form is $x^2 y'' + x(1 + x)y' + xy = 0$. So, $p_0 = 1$, $q_0 = 0$ and the indicial polynomial is r^2 with roots

$r_1 = r_2 = 0$. We are in Case II of Frobenius Theorem II. First, there is a solution of the form $y_1 = \sum_{n=0}^{\infty} a_n x^n$ with $a_0 = 1$. Using the techniques of Section 11.5, we see that the recursion formula in this case is

$$a_{n+1} = -a_n/(n+1), \quad n = 0, 1, 2, \cdots$$

and a solution is

$$y_1 = \sum_{n=0}^{\infty} (-1)^n x^n / n! = e^{-x}$$

We have by the Wronskian Reduction Method of Problem 6, Section 3.7, that $e^{-x}y_2' + e^{-x}y_2 = e^{-x}/x$, where the right-hand side is $\exp[-\int^x a(s)ds]$, $a(s) = s^{-1} + 1$, for the normalized ODE, $y'' + x^{-1}(1+x)y' + x^{-1}y = 0$. So, $y_2' + y_2 = 1/x$, or $(y_2 e^x)' = e^x/x$. We have as a second independent solution

$$y_2 = e^{-x} \int^x (e^s/s)ds$$

Since the integral can't be expressed in terms of elementary functions, we use power series techniques. Expanding e^s in its Taylor series about $x_0 = 0$, dividing each term by s, and integrating term by term, we have that

$$y_2 = e^{-x}[\ln x + x + x^2/4 + \cdots + x^n/(n!n) + \cdots] = y_1(x)[\ln x + x + x^2/4 + \cdots + x^n/(n!n) + \cdots]$$

(c). The ODE is $xy'' - xy' + y = 0$. Here the ODE in standard form is $x^2 y'' - x^2 y' + xy = 0$, and the indicial polynomial is $r^2 - r$ and $r_1 = 1$, $r_2 = 0$. We are in Case III of Frobenius Theorem II. We may use the methods of Section 11.5 to find a solution $y_1 = \sum_{n=0}^{\infty} a_n x^{n+1}$, $a_0 = 1$. The recursion formula is

$$a_{n+1} = \frac{(1-n)a_n}{n(n+1)}, \quad n \geq 0$$

So $a_n = 0$ for $n \geq 1$ and

$$y_1 = x$$

is a solution. We have by the Wronskian Reduction Method of Problem 6, Section 3.7 that $xy_2' - y_2 = e^x$, $y_2' - y_2/x = e^x/x$, $(y_2/x)' = e^x/x^2$ and

$$y_2 = x \int^x e^s/s^2 ds = x[-1/x + \ln x + \cdots + x^{n-1}/(n!(n-1)) + \cdots]$$

where we replaced e^s by its Maclaurin series, divided each term by s^2, and then integrated term by term.

3. **(a).** Let $y(x)$ be a solution of Bessel's equation of order p: $x^2 y''(x) + xy'(x) + (x^2 - p^2)y(x) = 0$. Let $w(z) = z^{-c}y(az^b)$. Then

$$w' = -\frac{cw}{z} + abz^{b-c-1}y'$$

$$w'' = \frac{c}{z^2}w - \frac{c}{z}(-\frac{c}{z}w + abz^{b-c-1}y') + ab(b-c-1)z^{b-c-2}y' + a^2 b^2 z^{2b-c-2}y''$$

A lengthy calculation then shows that w satisfies the equation $z^2 w'' + (2c+1)zw' + [a^2 b^2 z^{2b} + (c^2 - p^2 b^2)]w = 0$. Solutions of the w-equation can be expressed in terms of $z^{-c} J_p(az^b)$ and $z^{-c} Y_p(az^b)$ [or $z^{-c} J_{-p}(az^b)$ if p is not an integer].

(c). Comparing the given equation with the w-equation of **(a)**, we have that $c = -1/2$, $b = 2$, $a = 1/2$, $p = 1/(4\sqrt{2})$. So, $y = x^{1/2}[c_1 J_p(x^2/2) + c_2 J_{-p}(x^2/2)]$, $p = 1/(4\sqrt{2})$, $x > 0$.

5.   Group problem.

11.8 Steady Temperatures in Spheres and Cylinders

Comments

Finally, we find out why Bessel functions and Legendre polynomials were introduced in Chapter 11. It turns out that the wave, heat, and potential PDEs can be separated into individual coordinates only by using cylindrical or spherical coordinates when the bounding surfaces have those respective shapes. Moreover, one or more of the resulting ODEs will almost always be a Bessel equation, a Legendre equation, or some other nonconstant-coefficient second order linear ODE like those studied in Chapter 11. We have already seen in Section 10.9 that an Euler ODE comes up in solving a heat problem in a disk. In this section, we concentrate on just two models, steady temperatures in a cylinder immersed in ice water (except at one end), and steady temperatures in a spherical solid with prescribed temperatures on the boundary. Extensive use is made of the results in Sections 11.3, 4, 6, 7 on Legendre, Euler, and Bessel equations.

1.   We want to find a formula for the steady temperature in a unit ball with the given surface temperature $f(\phi)$. We follow the steps given in the text.

    (a). Because $f = \cos 2\phi - \sin^2 \phi = 3\cos^2 \phi - 2$, we have $g(x) = f(\arccos x) = 3x^2 - 2$. Using formulas (12) and (15), we have

$$u = \sum_{n=0}^{\infty} A_n \rho^n P_n(\cos \phi)$$

$$A_n = \frac{2n+1}{2} \int_{-1}^{1} g(x) P_n(x)\, dx = \begin{cases} -1 & n = 0 \\ 2 & n = 2 \\ 0 & \text{otherwise} \end{cases}$$

So, the steady-state temperature in a ball of unit radius is $u(\rho, \phi, \theta) = -1 + 2P_2(\cos \phi)\rho^2$.

    (c). Because $f = |\cos \phi|$, $g(x) = f(\arccos x) = |x|$. So

$$A_n = \frac{2n+1}{2} \int_{-1}^{1} g(x) P_n(x)\, dx = \frac{2n+1}{2} \int_{-1}^{1} |x| P_n(x)\, dx$$

$$= \frac{2n+1}{2} \left(\int_{0}^{1} x P_n(x)\, dx - \int_{-1}^{0} x P_n(x)\, dx \right)$$

Note from Table 11.3.1 that $n P_n = x P'_n - P'_{n-1}$, so we have

$$n \int_0^1 x P_n \, dx = \int_0^1 (x^2 P'_n - x P'_{n-1}) \, dx = x^2 P_n \Big|_0^1 - 2 \int_0^1 x P_n \, dx - x P_{n-1} \Big|_0^1 + \int_0^1 P_{n-1} \, dx$$

$$= P_n(1) - P_{n-1}(1) - 2 \int_0^1 x P_n \, dx + \int_0^1 P_{n-1} \, dx$$

$$= -2 \int_0^1 x P_n \, dx + \int_0^1 P_{n-1} \, dx$$

since $P_k(1) = 1$ for all k. Since $\int_0^1 P_k(x) \, dx = P_{k-1}(0)/(k+1)$, we have

$$\int_0^1 x P_n \, dx = \frac{1}{n+2} \int_0^1 P_{n-1} \, dx = \frac{P_{n-2}(0)}{n(n+2)}$$

Similarly, we can obtain

$$\int_{-1}^0 x P_n \, dx = \frac{1}{n+2} \int_{-1}^0 P_{n-1} \, dx = \frac{(-1)^{n-1} P_{n-2}(0)}{n(n+2)}$$

It is easy to calculate the coefficients A_0 and A_1 directly: $A_0 = 1/2$, $A_1 = 0$. For $n \geq 2$, we have that

$$A_n = \frac{2n+1}{2} \left(\frac{P_{n-2}(0)}{n(n+2)} + \frac{(-1)^{n-2} P_{n-2}(0)}{n(n+2)} \right) = \frac{(2n+1)(1+(-1)^{n-2})}{n(n+2)}$$

So, the steady-state temperature in a ball of unit radius is

$$u(\rho, \phi, \theta) = \frac{1}{2} + \sum_{n=2}^\infty \frac{(2n+1)(1+(-1)^{n-2}) P_{n-2}(0)}{n(n+2)} P_n(\cos \phi) \rho^n$$

3. **(a).** We are to solve boundary problem (16) in a cylinder if $f(r) = 1$, $0 \geq r \geq 1$. If $f(r) = 1$, $0 \leq r \leq 1$, then with x_n, $n = 1, 2, \ldots$, denoting the consecutive positive zeros of $J_0(x)$, we have from (20) that

$$A_n = \frac{2}{[J_1(x_n)]^2 \sinh(x_n a)} \int_0^1 r J_0(x_n r) \, dr = \frac{2}{[J_1(x_n)]^2 \sinh(x_n a) k_n^2} \int_0^1 (r J_1(x_n r))' \, dr$$

$$= \frac{2}{[J_1(x_n)]^2 \sinh(x_n a)) x_n^2} (r J_1(x_n r)) \Big|_0^1 = \frac{2}{J_1(x_n) \sinh(x_n a)) x_n^2}$$

where we have used the Bessel identity, $(x J_1(x))' = x J_0(x)$. So

$$u(r, z) = \sum_{n=1}^\infty \frac{2}{x_n^2} \cdot \frac{\sinh x_n (a - z)}{\sinh(x_n a)} \cdot \frac{J_0(x_n r)}{J_1(x_n)}$$

5. In a Neumann problem $\nabla^2 u = 0$ inside a region G and $\partial u / \partial n$ is a prescribed funtion of position on the boundary of G.

(a). Suppose that u_1 and u_2 are solutions and set $w = u_1 - u_2$. Then $\nabla^2 w = 0$ in G and $\partial w / \partial n = 0$ on ∂G. Let G^* consist of G together with its boundary. Since $\nabla(w^2) = 2w \nabla w$ and $\nabla w \cdot n = \partial w / \partial n = 0$, using the planar Divergence Theorem (Theorem B.5.13) on w^2, we have

$$\int_{\partial G} \nabla(w^2) \cdot n \, ds = \int_{G^*} \text{div}(\nabla w^2) \, dA = 0$$

On the other hand, since $\nabla^2 w = 0$,

$$\int_{G^*} \text{div}(\nabla w^2) \, dA = \int_{G^*} 2[w\nabla^2 w + ||\nabla w||^2] \, dA = 2 \int_{G^*} ||\nabla w^2|| \, dA = 0$$

So $\nabla w = \mathbf{i} \partial w / \partial x + \mathbf{j} \partial w / \partial y$ is the zero vector throughout G^*. That means that $w(x, y)$ does not change as x and y change. So $w = $ constant on G^*, and it follows that $u_1 = u_2 +$ constant.

(c). Suppose G is the unit disk in the plane and $f(\theta) = \sin \theta$ on the perimeter ∂G, while $u = 0$ if $\theta = 0$ and $r = 1$. As in formula (10) in Section 10.9, we have the following solution form:

$$u(r, \theta) = \frac{A_0}{2} + \sum_{n=1}^{\infty} r^n (A_n \cos n\theta + B_n \sin n\theta)$$

where A_n and B_n are arbitrary. We will determine A_n and B_n by imposing the boundary condition $\partial u / \partial n = \sin \theta$ on ∂G. That is

$$\frac{\partial u}{\partial r}(1, \theta) = \sin \theta = \sum_{n=1}^{\infty} [n A_n \cos n\theta + n B_n \sin n\theta]$$

We know that $A_n = 0$, $n = 1, 2, \ldots$, $B_1 = 1$, $B_2 = 0$, $n = 2, 3, \ldots$ and $u(r, \theta) = A_0/2 + r \sin \theta$. Since $u = 0$ when $\theta = 0$, $r = 1$, we must have $A_0 = 0$, so $u = r \sin \theta$.

APPENDIX A

Basic Theory of Initial Value Problems

Appendices A.1–A.4 discuss and verify the results of the Existence and Uniqueness Theorem (Theorem 2.1.1), the Extension Theorem (Theorem 2.2.1) and the Sensitivity/Continuity part of the Fundamental Theorem (Theorem 2.3.3). Appendix A.1 looks at Uniqueness, Appendix A.2 at Existence, Appendix A.3 at Extension, and Appendix A.4 at Sensitivity/Continuity.

A.1 Uniqueness

1. **(a).** The function $f = |y|$ is not differentiable at $y = 0$ since $\lim_{h \to 0}(f(h) - f(0))/h = \lim_{h \to 0}((|h| - 0)/h) = \pm 1$, depending upon whether h goes to zero through positive or negative values. So f doesn't satisfy the hypotheses of Theorem 2.1.1.

   (c). The proof of the Uniqueness Principle in the textbook actually works when f satisfies a Lipschitz condition. From the Triangle Inequality (Section 7.2) we have that $||y_1| - |y_2|| \leq |y_1 - y_2|$, for all y_1, y_2. So $f(y) = |y|$ satisfies a Lipschitz condition [see (4) in the text] with $L = 1$; it follows that the IVP of part **(b)** has a unique solution.

A.2 The Picard Process for Solving an Initial Value Problem

1. We want to find the Picard iterates for $y' = -y$, $y(0) = 1$.

   (a). Using iteration scheme (5), we have the Picard iterates $y_0(t) = 1$, all t

   $$y_1(t) = 1 + \int_0^t (-y_0(s))\, ds = 1 + \int_0^t (-1)\, ds = 1 - t$$

   $$y_2(t) = 1 + \int_0^t (-y_1(s))\, ds = 1 + \int_0^t -(1 - s)\, ds = 1 - t + t^2/2$$

   $$y_3(t) = 1 + \int_0^t (-y_2(s))\, ds = 1 + \int_0^t -(1 - s + s^2/2)\, ds = 1 - t + t^2/2 - t^3/6$$

   These iterates are defined for all t.

(c). Following the pattern of y_0, y_1, y_2, y_3, we guess that the n-th Picard iterate is given by $y_n(t) = \sum_{j=0}^{n}(-1)^j t^j/j!$. This is proved by induction on n: The formula holds for $n = 1$ (see part **(a)**). Suppose it holds for n (the induction hypothesis). We shall show it holds for $n + 1$. From the iteration scheme (5) we have

$$y_{n+1} = 1 + \int_0^t (-y_n(s))\, ds = 1 + \int_0^t \left(-\sum_{j=0}^{n} \frac{(-1)^j s^j}{j!}\right) ds = 1 - \sum_{j=0}^{n} \int_0^t \frac{(-1)^j s^j}{j!}\, ds$$

$$= 1 + \sum_{j=0}^{n} \frac{(-1)^{j+1} t^{j+1}}{(j+1)!} = \sum_{j=0}^{n+1} \frac{(-1)^j t^j}{j!}$$

where we have used the induction hypothesis and reindexed a sum. So the formula holds for $n + 1$, if it holds for n. By induction, the formula holds for all n, and all t.

(e). The Taylor series for e^{-t} is $\sum_{j=0}^{\infty}(-1)^j t^j/j!$ about $t_0 = 0$ since if $f(t) = e^{-t}$ then $f^{(j)}(0) = (-1)^j$, and the Taylor series for $f(t)$ is $\sum_{j=0}^{\infty} f^{(j)} t^j/j!$. So $y_n(t)$ is just the partial sum $\sum_{j=0}^{n}(-1)^j t^j/j!$ of the Taylor series for e^{-t}. In this problem, Picard iteration constructs the Taylor series for the solution by adding one term to the series at each iteration.

3. The IVP is $y' = y^2$, $y(0) = 1$.

(a). Using the iteration scheme (5) to calculate the Picard iterates, we have

$$y_0(t) = 1,$$

$$y_1(t) = 1 + \int_0^t y_0^2(s)\, ds = 1 + \int_0^t 1\, ds = 1 + t$$

$$y_2(t) = 1 + \int_0^t y_1^2(s)\, ds = 1 + \int_0^t (1+s)^2 ds = 1 + t + t^2 + t^3/3$$

$$y_3(t) = 1 + \int_0^t y_2^2(s)\, ds = 1 + \int_0^t (1 + s + s^2 + s^3/3)^2\, ds$$

$$= 1 + t + t^2 + t^3 + 2t^4/3 + t^5/3 + t^6/9 + t^7/63$$

All are defined for all t.

(c). Certainly, $y_1(t) = 1 + t$ is a polynomial of degree 1. Assume that $y_n(t)$ is a polynomial of degree $2^n - 1$ (the induction hypothesis). To show that $y_{n+1}(t)$ is a polynomial of degree $2^{n+1} - 1$, we note that

$$y_{n+1}(t) = 1 + \int_0^t y_n^2(s)\, ds = 1 + \int_0^t (c_0 + \cdots + c_k s^{2^n-1})^2\, ds$$

$$= 1 + \int_0^t (c_0^2 + \cdots + c_k^2 s^{2^{n+1}-2})\, ds = \cdots + c_k^2(2^{n+1} - 1)^{-1} t^{2^{n+1}-1}$$

where we have set $y_n(t) = c_0 + \cdots + c_k t^{2^n-1}$. So $y_{n+1}(t)$ is a polynomial of the desired degree and the proof is complete.

5. In this case we have $y' = -y$, $y(0) = 1$, but we take $y_0 = \sin t$ as the zeroth Picard iterate:

$$y_0(t) = \sin t$$

$$y_1(t) = 1 + \int_0^t (-y_0(s))\, ds = 1 + \int_0^t (-\sin s)\, ds = \cos t,$$

$$y_2(t) = 1 + \int_0^t (-y_1(s))\, ds = 1 + \int_0^t -\cos s\, ds = 1 - \sin t$$

If we continue this process, being careful to separate the terms as above, we see that each $y_n(t)$ is the sum of two terms, the first being the partial sum $\sum_{j=0}^{n-1} (-1)^j t^j/j!$ of the Taylor series of e^{-t}, while the magnitude of the second term is the magnitude of the difference between $\pm \sin t$ [or $\pm \cos t$] and its corresponding Taylor series through the term t^{n-1}. Since the latter Taylor series tends to $\sin t$ [or $\cos t$] as $n \to \infty$, these terms tend to zero as $n \to \infty$ and the sequence of iterates converges to e^{-t}, even with the absurd choice of $y_0(t) = \sin t$.

7. The IVP is $y' = \sin(t^3 + ty^5)$, $y(0) = 1$. We have $y_0(t) = 1$, $y_1(t) = 1 + \int_0^t \sin(s^3 + s)\, ds$, which cannot be expressed in terms of elementary functions. In general, the use of the Picard scheme is limited by the difficulty of carrying out the integration of any but the simplest of functions. This does not mean that the integrals fail to exist, only that they cannot be written in terms of elementary functions.

9. **(a).** Let ε be any positive number. We must find a positive integer N, which may depend upon ε, so that for all integers $n \geq N$, $|(\sin nt)/n| < \varepsilon$, for all t in some interval I. Since $|(\sin nt)/n| \leq 1/n$, we see that we may take for N any integer larger than $1/\varepsilon$. Then if $n \geq N$, we have

$$|\sin nt/n| \leq \frac{1}{n} \leq \frac{1}{N} < \frac{1}{1/\varepsilon} = \varepsilon$$

So $\{(\sin nt)/n\}_1^\infty$ converges uniformly to the zero function on any interval I.

A.3 Extension of Solutions

---

1. **(a).** The variables in the ODE $y' = -y^3$ may be separated and the ODE solved (using the initial condition $y(0) = 1$) to obtain $y = (2t+1)^{-1/2}$. This solution is defined for $t > -1/2$. As t decreases to $-1/2$, $y(t)$ increases to $+\infty$, while as $t \to +\infty$, $y(t)$ decreases to 0.

3. The IVP is $y' + p(t)y = q(t)$, $y(t_0) = y_0$, where p and q are continuous on an interval I. The function $P(t) = \int_{t_0}^t p(s)\, ds$ is continuous for all t in I since $p(t)$ is continuous for all t in I. So $e^{P(t)}$ and $\int_{t_0}^t e^{P(s)}q(s)\, ds$ are continuous for all t in I, the latter since $q(t)$ is continuous for all t in I. If $p(t)$ and $q(t)$ are continuous for all t in I, the solution formula (3) in Section 1.3 tells us that $y(t) = e^{P(t_0)}y_0 e^{-P(t)} + e^{-P(t)}\int_{t_0}^t e^{P(s)}q(s)\, ds$, so $y(t)$ is defined for all t. There is no "finite escape time" inside I.

A.4 Sensitivity of Solutions to the Data

1. The two IVPs to be compared are $y' = y/10$, $y(0) = a$, and $y' = y/10 + ye^{-y^2}$, $y(0) = \tilde{a}$, with respective solutions $y(t)$ and $\tilde{y}(t)$. Since the number M in the Perturbation Estimate is an upper bound on the "perturbation" term ye^{y^2}, we find the maximum value M of that term by equating its derivative to 0 and finding $y = 2^{-1/2}$. So we can take $M = 2^{-1/2}e^{-1/2}$: for simplicity we take $M = 1/2$, which is slightly larger. The number L in the estimate (Theorem A.4.2) is an upper bound for $|\partial(y/10)/\partial y| = 1/10$, so $L = 1/10$. Since $t_0 = 0$, the inequality (11) becomes

$$|y(t) - \tilde{y}(t)| \leq |a - \tilde{a}| e^{t/10} + 5 \left(e^{t/10} - 1\right), \qquad 0 \leq t$$

The coefficient 5 is not unique.

3. Now $y(t)$ and $\tilde{y}(t)$ are both solutions of the ODE $y' = f(t, y)$, so $y'(t) = f(t, y(t))$ and $\tilde{y}'(t) = f(t, \tilde{y}(t))$, for all $a \leq t \leq b$. Integrating these equations over the interval $a \leq s \leq t$, we see that

$$y(t) = y(a) + \int_a^t f(s, y(s)) \, ds, \quad \text{and} \quad \tilde{y}(t) = \tilde{y}(a) + \int_a^t f(s, \tilde{y}(s)) \, ds$$

for all $a \leq t \leq b$. Without loss of generality, suppose that $y(a) > \tilde{y}(a)$. Then from the Existence and Uniqueness Theorem (Theorem 2.1.1) we see that $y(t) > \tilde{y}(t)$, so $|y(t) - \tilde{y}(t)| = y(t) - \tilde{y}(t)$, for all $a \leq t \leq b$. Subtracting the second integral equation from the first, we see that

$$y(t) - \tilde{y}(t) = y(a) - \tilde{y}(a) + \int_a^t \{f(s, y(s)) - f(s, \tilde{y}(s))\} \, ds$$

and from the Mean Value Theorem (Theorem B.5.4) it follows that

$$y(t) - \tilde{y}(t) = y(a) - \tilde{y}(a) + \int_a^t f_y(s, c(s))(y(s) - \tilde{y}(s)) \, ds \qquad \text{(i)}$$

where c is a value (which may depend on s) with $\tilde{y}(s) < c(s) < y(s)$.

We are now ready to address the bulleted items in turn:

- If $K = 0$, then $f_y(t, y) \geq 0$ for all (t, y) in R. We have

$$\int_a^t f_y(s, c(s))(y(s) - \tilde{y}(s)) \, ds \geq 0, \quad \text{for all } a \leq t \leq b$$

so from (i) above we see that

$$y(t) - \tilde{y}(t) \geq y(a) - \tilde{y}(a), \quad \text{for all } a \leq t \leq b$$

Since $|y(t) - \bar{y}(t)| = y(t) - \bar{y}(t)$, for all $a \le t \le b$, the desired inequality follows. If $L = 0$, then $f_y(t, y) \le 0$, for all (t, y) in R, and the asserted inequality follows in a similar fashion.

- Now assume that $0 < K \le f_y(t, y) \le L$ for all (t, y) in R. We have

$$K \int_a^t (y(s) - \bar{y}(s)) \, ds \le \int_a^t f_y(s, c(s))(y(s) - \bar{y}(s)) \, ds \le L \int_a^t (y(s) - \bar{y}(s)) \, ds$$

for all $a \le t \le b$. If we put $z(t) = y(t) - \bar{y}(t)$, then from (i) above we have that

$$K \int_a^t z(s) \, ds \le z(t) - z(a) \le L \int_a^t z(s) \, ds, \quad \text{for all } a \le t \le b \qquad \text{(ii)}$$

Following the proof of the Basic Estimate (Theorem A.4.1), we put $w(t) = \int_a^t z(s) \, ds$, and observe that $w'(t) = z(t)$ and so the inequality above becomes

$$Kw \le w' - z(a) \le Lw \qquad \text{(iii)}$$

Working with the top half of inequality (iii), we have that

$$w' - Lw - z(a) \le 0, \quad \text{for } a \le t \le b$$

which, after multiplying through by e^{-Lt} becomes

$$\left(we^{-Lt} + \frac{z(a)}{L} e^{-Lt} \right)' \le 0, \quad \text{for } a \le t \le b$$

The function $[w - z(a)/L]e^{-Lt}$ is a nonincreasing function, and so

$$\left(w + \frac{z(a)}{L} \right) e^{-Lt} \le \left(w(a) + \frac{z(a)}{L} \right) e^{-La} \qquad \text{(iv)}$$

or [since $w(a) = 0$]

$$Lw + z(a) \le z(a)e^{L(t-a)}, \quad a \le t \le b$$

But from (iii), $z(t) \le Lw + z(a)$, and so the desired inequality holds:

$$z(t) \le z(a)e^{L(t-a)}, \quad a \le t \le b$$

To show the other half of the asserted inequality, write the bottom part of inequality (iii) as

$$\left(we^{-Kt} + \frac{z(a)}{K} e^{-Kt} \right)' \ge 0, \quad \text{for all } a \le t \le b$$

The function $(w + z(a)/K)e^{-Kt}$ is nondecreasing, so

$$(w(t) + z(a)/K) e^{-Kt} \ge (w(a) + z(a)/K) e^{-Ka} \qquad \text{(v)}$$

or [since $w(a) = 0$]

$$Kw + z(a) \ge z(a)e^{K(t-a)}, \quad a \le t \le b$$

But from (iii), $z(t) \geq Kw + z(a)$, so the desired inequality holds:

$$z(a)e^{K(t-a)} \leq z(t), \quad a \leq t \leq b$$

The above argument fails when $K < 0$ or $L < 0$ because, when the inequalities (iv) or (v) are multiplied by K or L, the resulting inequality goes the wrong way.

- We treat only the case where $L < 0$; the case where $K < 0$ is treated similarly. From the top half of (ii) we have that

$$z(t) \leq z(a) + L \int_a^t z(s)\, ds, \quad a \leq t \leq b$$

where $z(t) = y(t) - \bar{y}(t) \geq 0$, for all $a \leq t \leq b$. Iterating the above inequality, we have that

$$z(t) \leq z(a) + L \int_a^t \left\{ z(a) + L \int_a^s z(r)\, dr \right\} ds$$

$$\leq z(a) + z(a)L(t-a) + L^2 \int_a^t \left\{ \int_a^s z(r)\, dr \right\} ds$$

Changing the order of integration in the iterated integral, we obtain

$$z(t) \leq z(a) + z(a)L(t-a) + L^2 \int_a^t z(r) \left\{ \int_r^t ds \right\} dr$$

$$\leq z(a) + z(a)L(t-a) + L^2 \int_a^t (t-r)z(r)\, dr$$

Continuing the iteration in this manner we obtain after n iterations

$$z(t) \leq z(a) + z(a)L(t-a) + z(a)\frac{L^2(t-a)^2}{2!} + \cdots$$

$$+ L^{n+1} \int_a^t \frac{(t-r)^n}{n!} z(r)\, dr$$

Since the last term in the estimate above goes to zero for any choice of t in the interval $a \leq t \leq b$, it follows that

$$z(t) \leq z(a) \left\{ 1 + L(t-a) + \frac{L^2(t-a)^2}{2!} + \cdots + \frac{L^n(t-a)^n}{n!} + \cdots \right\}$$

$$\leq z(a)e^{L(t-a)}$$

as was asserted.

- Observe that $f_y(t, y) = (-1/10)y/(1+y)$, and so

$$-\frac{1}{10} \leq f_y(t, y) \leq 0, \quad \text{for } t \geq 0, \ y \geq 0$$

We have the estimates

$$|y(0) - \bar{y}(0)|e^{-t/10} \leq |y(t) - \bar{y}(t)| \leq |y(a) - \bar{y}(a)|$$

for $t \geq 0$, and any two solutions $y(t)$, $\bar{y}(t)$ of the ODE $y' = (1/10)(-y + \ln(1 + y))$ which lie in the closed region $t \geq 0$, $y \geq 0$. See Fig. 3, Graph 1 for the graphs of the two solutions of the ODE with data $y(0) = 4$, and $\bar{y}(0) = 1$. It is evident that $|y(t) - \bar{y}(t)| < 3$, for all $0 \leq t \leq 40$. Direct measurement shows that $|y(10) - \bar{y}(10)| \geq 3e^{-1} \approx 1.104$, and that $|y(20) - \bar{y}(20)| \geq 3e^{-2} \approx 0.406$, and that $|y(40) - \bar{y}(40)| \geq 3e^{-4} \approx 0.055$.

- Since $f(t, y) = \sin t - (1/10)(0.5 \cos y + y)$, we see that

$$f_y(t, y) = (-1 + 0.5 \sin y)/10$$

so

$$-0.15 \leq f_y(t, y) \leq -0.05, \quad \text{for all } t \text{ and } y$$

It follows from the estimate proven above that

$$10e^{-0.15t} \leq |y(t) - \bar{y}(t)| \leq 10e^{-0.05t}, \quad \text{for all } t$$

for the given solutions $y(t)$ and $\bar{y}(t)$. See Fig. 3, Graph 2.

Problem 3, Graph 1.

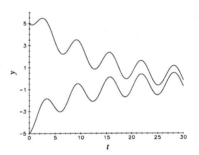

Problem 3, Graph 2.

Notes

Notes

Notes

Notes

Notes

Notes

Notes

Notes

Notes